Foster Wilson

Cotton Carders' Companion

Foster Wilson

Cotton Carders' Companion

ISBN/EAN: 9783337061036

Printed in Europe, USA, Canada, Australia, Japan

Cover: Foto ©berggeist007 / pixelio.de

More available books at **www.hansebooks.com**

COTTON CARDERS' COMPANION.

IN WHICH IS GIVEN A DESCRIPTION OF THE

Manner of Picking, Baling, Marketing, Opening and Carding Cotton.

TO WHICH IS ADDED A LIST OF

VALUABLE TABLES, RULES AND RECEIPTS.

By FOSTER WILSON, Lowell, Mass.

"Not slothful in business; fervent in spirit, serving the Lord."—ROM. XII: 11.

"A blessing now,—a curse no more,
   Since He whose name we breathe with awe,
The coarse mechanic vesture wore,—
   A poor man toiling with the poor,
     In labor as in prayer, fulfilling the same law."
                              —WHITTIER.

LOWELL, MASS.:
STONE, HUSE & CO., BOOK AND JOB PRINTERS, No. 130 CENTRAL ST.
1876.

INTRODUCTION.

The author makes his lowest bow to his patrons as he introduces the COTTON CARDERS' COMPANION. Many works have been written on cotton manufacturing—written mostly by gentlemen and scholars. The author has read them with great profit, and yet he has many times felt the need of just such a work as this. His work is not designed to take the place of any that have preceded it, but is a sort of little particular supplement. Most works on this subject take in the whole of manufacturing. This only carding. Many deal largely in theories. This in practice. Some are general. This is particular. This little work will come to the notice of the learned, as well as laboring men. The former will please bear in mind that the writer of it is himself a laboring man; and that it is designed particularly to aid his own class in the successful prosecution of their vocation; though it contains some ideas worthy the notice of any engaged in the manufacture of cotton goods. And though the style is simple, he trusts the matter may be profitable. If a book is good, it does not need a long preface. If it is dry, a long preface only makes it drier.

Very respectfully,

FOSTER WILSON.

INDEX.

CHAPTER I.
Cotton, and the Machinery used in its Manufacture, ... 5

CHAPTER II.
Opening, or Picking, Cotton, ... 13

CHAPTER III.
Carding, ... 34

CHAPTER IV.
Drawing, ... 76

CHAPTER V.
Speeders, ... 96

CHAPTER VI.
Miscellany, ... 119

PARTICULAR INDEX.

Re-working Waste, ... 17
Grinding, ... 38
Amount of Work for a Card, ... 48
Stripping Cards, ... 55
Screens, ... 59
Laying Drawing in Railway Boxes, ... 79
Regulating Tension, ... 100
Progressive Doublings and **Drawing**, ... 109
Lowell Speeder Tables, ... 114
Hank Table, ... 119
Twist Table, ... 124
Hygrometer Table, ... 132
Belt Table, ... 140
Perpetual Calendar, ... 142
Weights of Various Articles, ... 149
Receipts, ... 150

Entered according to Act of Congress, in the year 1876, by
FOSTER WILSON,
In the Office of the Librarian of Congress, at Washington.

CHAPTER I.

Cotton, and the Machinery used in its Manufacture.

GENERAL REMARKS — COTTON — WHAT IS IN IT — HOW IT GETS THERE — A CARDER'S BUSINESS — PICKING COTTON — DRYING COTTON — GINNING COTTON — BALING COTTON — TRANSPORTING COTTON — AFFECTED BY THE ATMOSPHERE AND ELECTRICITY — MACHINERY AND IMPROVEMENTS — WHAT IT SHOULD BE, AND HOW KEPT SO.

IN a work of this kind it seems appropriate to begin at the bottom. When a physician is called to a sick man, the first thing he does is to find out the nature of the disease, so that he may be able to treat it successfully. When we are about taking a journey to a distant place, we first acquaint ourselves with the condition of the road that leads there, that we may know what kind of conveyance is best adapted to it. If we have a material to work upon, it is well to inform ourselves as to its nature and general character, that we may provide ourselves with the proper implements. If the material is cotton the first thing to do is to find out what cotton is, what is its condition, how it came to be in such a condition, and what is necessary to work such changes as we desire.

We find cotton to be the fibre of a plant; and if there was nothing else but good, nice, straight fibre, we should find it comparatively easy to convert it into yarn and cloth. But unfortunately for us, good long, straight fibre is mixed up with other things, and such other things as we do not want, and must be rid of in order to utilize the fibre. Cotton as it comes to us has dead leaf, sand, seeds, blights, and what we call nits and gins, and the fibre is considerably curled up. The business of a cotton carder is to rid it of all extraneous matter, lay the fibres parallel, and form them into a thread called roving.

It may be interesting and profitable to examine into the causes of so much foreign matter being mixed up with the fibre of cotton; and to do this we must take a look at it in the field where it grows, watch the process and progress of picking, preparing for market and its transportation.

Cotton grows in what it would be natural for us to call pods, but the planters call them bolls. When it is ripe the bolls open; the fibre is curled up round the seed. This is a provision of nature for its own propagation. But at this point art lays hold of it and appropriates it. The bolls begin to open the latter part of July or first of August, according to the season. When a sufficient number of bolls have opened the pickers commence their work; each is provided with a bag suspended from the neck, so that both hands may be used. The contents of all the open bolls are transferred to the bag. The bolls that have been open longest have some of them shed their contents out on to the ground, or they have been beaten out by the action of the elements, but they are all gathered up by the picker; hence we find dirt in cotton. The bolls do not open all at once, and by the time the pickers have passed over the fields once, another set are open, and the hands pass over it a second and third time, and thus it continues sometimes as late as December.

During this time there is usually some frost which stops the growth of the plant and kills the leaves and the leaves of weeds and other vegetation contiguous to the cotton plant. These leaves are soon dry, crumble up, get mixed with the cotton, and find their way into the picker's bag. When the frost comes and puts a stop to the growth there are some which are not ripe, and the green, frost-bitten boll dries up and turns black; the unripe cotton inside clings to it; this kind is also transferred to the picker's bag. So we find dead leaf and blighted cotton as well as dirt.

If the weather is cloudy and damp at the time of picking, the cotton is spread out to dry for a time. After it is dry it is taken to the gin-house to be ginned. The gin is a savage sort of machine, and it may be so worked as to injure the fibre; indeed cotton often bears unmistakable evidence of injury received in ginning. The savage gin referred to is called the "saw gin," which is the only kind the writer has ever seen, and those at rest, and he is not qualified to give an intelligible description of them; yet from what he does know of the character of cotton and the effect of machinery upon its fibre, he is able to form some idea of how injury may be done in the process.

The plan is to draw the fibre through a wire frame or rack by means of saw teeth. Let the reader picture to his mind a number of smallish circular saws, fixed on a shaft, or arbor, side by side, with the teeth just projecting through the bars of a wire rack; behind the saws a revolving brush to clear them. Thus much the writer has seen, and the rest he must himself imagine. We will suppose that the machine is set in motion— saws revolving with considerable velocity, and the brush in an opposite direction to clear the saws, the operator placing the cotton in the wire rack; the saw teeth lay hold of it and draw the fibre through, leaving the seeds behind, the brush cleaning the saws and passing the cotton along towards the delivery. Now, suppose everything is in order and the cotton properly fed to the machine, we can conceive how the work may be very nicely done; but suppose the machine to be out of order, the brushes too far off, or they may have become worn and consequently ineffective, or the operator may crowd the machine. In either of the above ways, and perhaps in others, the machine may work badly, get loaded up, the cotton pass through between the bars of the wire rack several times on the points of the saw teeth, and get bruised and ground up, producing that peculiar little white nub, or nit, or nep; it is something familiar to carders, whatever its proper name may be, and after once being formed they are hard to get rid of, and consequently it is very important that cotton be properly ginned.

There is another kind of gin called the roller gin, less severe in its action upon the fibre, though slower in its operation. The object of ginning is to rid cotton of seeds. This it does in a good degree, though some find their way through the bars of the rack or the rolls in a roller gin, and hence we have some seeds in addition to all other substances that have been mentioned.

The next process after ginning is baling. This is done in about the same manner as farmers bale hay in New England—a press with a long wooden screw worked by man, or horse power, or both. The bales as put up by the planters are very much larger than they are when we get them. After it is baled it is taken to the landing on the river, or to the railroad station, as the case may be. At these places it often lies for days waiting for shipment; sometimes rains fall on it, and it frequently gets tumbled about in the dirt, or mud, or both, considerably before loading on the steamer or cars.

On arriving at market (the planters' market) it is inspected, assorted, rebaled under a hydraulic press; and here the bales are pressed harder, and made smaller for convenience of transportation. It is now ready for market. It goes through a similar carting and rolling about in the mud to that described above.

Southern mud is very adhesive and will stick to a cotton-bag as well as anything else, and does so stick. When this mud gets dry much of it works through the bag, and we find it in the shape of sand; so we have another ingredient in compound. It may be well to sum them up: First, dirt gathered with the cotton in the field; second, dead leaf; third, unripe or blighted cotton; fourth, nits made in ginning, in some cases; fifth, seeds; sixth, sand. When cotton is rebaled, it is frequently damp, and the tremendous pressure to which it is subjected renders it more difficult to relieve from these substances than it otherwise would be.

Cotton differs in one particular from any other vegetable fibre. It is this: each fibre is a little independent twisted thread, the twist so close that it cannot be seen with the naked eye, though it is discernible with a powerful glass. This property gives the fibres an attraction to each other, and an adhesiveness after being joined by extraneous twist, rendering it capable of a tenuity far beyond any other vegetable fibre. It is considerably affected by the atmosphere, and in working it is subjected to the continual changes which are taking place in that element. It will absorb five per cent. of its own weight in dampness in the atmosphere, under some circumstances.

It is also affected by electricity, though a nonconductor. Electricity is excited more or less by machinery through which cotton passes in its manufacture, and when the air is dry it is also a non-conductor. Under these circumstances the electricity remains where it is excited, and is frequently the cause of great annoyance to cotton carders. We used to say—" When the air is dry it is full of electricity." The fact is, when the air is dry it does not conduct the electricity away, and it remains where it is excited, attracting cotton fibres in all directions, drawing rolls up, and roving will not draw down, nor twist smooth. These facts should be borne in mind by cotton carders. The remedy for these evils will be considered, each under its appropriate head, as this work progresses.

We have now traced cotton from the boll to the bale. We have seen how other substances have got mixed up with it, and how hard they have been pressed into it. Our next business will be to consider how to get these foreign substances out of it, the fibres completely separated from them and laid parallel. It will be our object, in considering this subject, to point out the most convenient, and at the same time efficacious, manner of doing it.

The primitive old plan of doing this business was very simple, performed by hand, slow and tedious; so that when men began to multiply rapidly and began to have a desire to dress fashionably, the supply fell short of the demand. Then it was, some hundred years ago or more, that men in England began to turn their attention to the invention of machinery to assist in this branch of cotton manufacture, as it had been employed in spinning. The first attempts in this direction were rude and simple; but as one machine after another was constructed and applied to the purpose, improvements suggested themselves to the minds of the builders and operators; new machines were built, old ones were remodeled, and faults remedied from time to time, until a whole system of machinery was arranged to take cotton from the field and change it into cloth. And when the system was completed, improvements continued and are vigorously kept up until this present, and are likely to continue as long as the world stands. We have learned something, and yet we are learning. "Every day brings something new," is as true in cotton manufacturing as in anything else.

Sometimes enthusiastic inventors have a zeal with which unfortunately their knowledge does not keep pace, and we are offered some impracticable machinery. But good judgment is sure to come to the rescue, sooner or later, and we are enabled to discern between that which is right and proper and that which is not—keep the good and cast the bad away. We are sometimes troubled with poorly-built machinery. Cheapness is the great desideratum. This has led to a sharp competition and given us some machines that are not A No. 1. Accuracy of construction is very necessary in all cotton machinery, but nowhere more so than in that belonging to the carding department. Cast-iron pulleys work more satisfactorily when covered with leather; belts are not required to be so tight when they are so covered, as they are without such covering, thus making a saving not only in belts and belt fastenings, but a saving in shafting, hangers and oil. All kinds of shafting should be hung level and straight, and in new mills, where there is a liability of shrinkage of wood-work and settling of walls, it should be re-leveled as often as occasion may require. Every carder should insist on having thoroughly-built and accurate machinery; and if it is not so, he should report to the proper authorities, so that censure may come in the right place, if it comes at all. All kinds of machinery should be carefully watched and kept in repair; the first beginnings of disorder should be attended to. "A stitch in time saves nine" is true in this particular. Machinery should not be left to run alone—not even on loose pulleys—as there is a liability of trouble. A pulley may heat and throw fire, or it may stick and start a frame.

CHAPTER II.

Opening, or Picking, Cotton.

GENERAL REMARKS — CHANGES — PURPOSE AND PRINCIPLES OF OPENING AND PICKING COTTON — OLD METHOD — GRADUAL CHANGES — NAMES OF MACHINES — INVENTIONS — VARIATIONS IN COTTON — MIXING — THE BEST METHOD — REASONS GIVEN — WASTE, HOW IT WILL WORK ALONE — FIRST MACHINE — WHAT IT USED TO BE — A WHIPPER — WHAT SPEED TO RUN — THE CALVERT WILLOW — OLD STYLE — IMPROVEMENTS ON SAME — NEW MACHINERY — ADVANTAGES CLAIMED — PRODUCTS OF SAME — RELATIONS OF BEATERS AND ROLLERS — SPEED OF BEATERS — QUALIFICATIONS OF TERMS — DRAUGHT OF PICKERS — FAN DRAUGHT — SELF-REGULATING DOORS AND EVENERS — DOUBLE PICKING — SINGLE — REGULATING BY SOUND OF BEATERS — OLD PRACTICE OF WORKING WASTE — NEW ONE SUGGESTED — OBJECTIONS CONSIDERED — MAKING THE MOST OF EVERYTHING — INJURY OF FIBRE — WHAT DO WE MEAN BY IT — WHAT IT IS NOT — WHAT IS NECESSARY — HOW TO MANAGE — THE MODERN PICKER AT WORK — WHAT IT DOES, AND HOW — THE CREDIT OF THE MACHINE TO WHOM IT BELONGS — KITSON'S NEW OPENER: A DESCRIPTION OF — SPRINKLERS — A SUGGESTION.

PICKING as it used to be called was, and is now, the first process that cotton passes through in its manufacture; and though this process has been known in different sections, at different times, by various names, such as "whipping," "willowing," "picking" and "opening," and the names of the machines used for these purposes have been as varied as the names of processes, yet the purpose and principle remain the same. The purpose is to open out and disentangle the fibre, relieve from dirt as much as possible, and prepare it for the card. The principle is to subject it to the action of machinery of some kind, with more or less severity of whipping, scratching and pounding, to produce the desired result. Within the last twenty years, more or less, this branch of business has been made a department by itself by many large concerns and some small ones. It is common in these times to find in many places the "picker house" (as it is usually called), with its appurtenances, all independent of the carding, or any other room. Where this is the case, of course picking is no part of the carder's business. But there are still some places where this business is carried on in connection with the carding, and under the supervision of the carder, and consequently must be considered in this volume. With the change in style of doing this work, making it a department by itself, it becomes a question of more interest than formerly. More attention is paid to it now than used to be under the old system.

Many changes, and some improvements, have been made in machines; but as has been remarked, the purpose and principle remain the same. It is the manner that changes. The old method of separating the cotton fibre from seeds and dirt was first to spread a quantity on a bench or platform, and whip it with willow sticks. This loosened out the fibre a little; then it was picked by hand; then carded between two flat hand cards. It is very probable that, as machinery began to take the place of some of the instruments used for these purposes, the machines took the names of instruments. Machines were first worked by hand-power, then horse-power, then water and steam.

In this light it is easy to conceive how machines used for loosening out and cleaning cotton fibres were first called a "willow" and a "picker," because they originally performed the work with willow sticks and by hand picking. "Scutch" was the name of an instrument used in England in old times to break flax with; so in that country what used to be called a "picker" with us, was by them called a "scutcher." We are getting nearer together now, and most people call this business opening and picking cotton.

It is not necessary, in a work of this kind, to trace out the names of inventors from those primitive times down to the present; neither would it be safe to undertake it, as many of the same inventions are claimed on both sides of the Atlantic. It is generally conceded, however, that a Scotchman by the name of Snodgrass was the inventor of the beater principle, though his original idea has been greatly enlarged upon, and still the work is going on. A sentence in the circular announcing this work, reads as follows: "The best method of putting cotton through the various processes through which it passes in preparation for spinning will be discussed at some length, each under its appropriate head, and reasons given." And the writer would invite particular attention to the last words of the sentence, "And reasons given."

The writer has had some experience in the preparation of cotton, and in some particulars he knows whereof he speaks. It is not guess-work, though he would not lay down a particular rule, to be governed by under all circumstances. Cotton varies so much in length, strength, fineness, smoothness and cleanness of fibre, that it would be very unwise to treat it all alike in preparation; but there are some general rules that are applicable to all kinds.

And under this head may be mentioned, mixing previous to working with machinery. In this particular it may; yea, it *should* all be treated alike. Men have their own peculiar notions as to the best course to pursue in mixing. The writer has tried various ways, under direction of parties for whom he has worked, some of them very cumbersome and laborious; and he has no fault to find with any of them. But there is one plan he has tried with entire satisfaction; while there may be many good ways, this one is considered the best. First, a large, airy room should be provided—the larger the better. The bales, after removing the bags, should be laid down close together on the edge. Proceed in this manner until the room is filled, all except a space near the first machine through which it is to pass. Second, card the bags, and place the cotton thus removed in a pile by itself. The reason for laying bales down edgewise instead of sidewise, is because the bales were packed in that way, and they pick up very much nicer and easier from the edge than they do from the side. They should be allowed to lie twenty-four hours or more after they are taken into the room, before any is used. The reason for this is, when the bag is first removed the cotton in the bale is damp, and this time is necessary for it to dry. In this manner the top of all the bales will dry down a certain distance, become kind of loose and light by exposure to dry air, and the cotton works easier.

Next comes "picking up cotton," in order to mix. It is impossible to make uniform work in the carding-room unless the cotton is well mixed in the picker-house, it varies so much in quality, even in the same grade; and it is often the case that two or more grades are worked together, so that, if all the long staple in a certain lot should be worked together, then the medium, and lastly the short, the ups and downs in the work that follows would be intolerable. To avoid this it must be mixed in the picker-room. It is thought by some that where there is double carding, and a large number of breaker cards run into one lap-head, the cotton is mixed in this way, and it is; but this plan can never be substituted for mixing in the picker-room. To secure good, uniform work, all kinds must be mixed together before it is worked at all. As the cotton is picked up, a layer from each bale some six, eight, or ten inches deep, as the case may be (enough to last one day), the waste from the carding-room and spinning-room—one day's waste, or all the waste made in these rooms in one day, that is calculated to be re-worked in all the machines through which the cotton passes—should be mixed with it; also, a portion of the cotton that has been carded from the bags; and the whole thrown in a pile near the first machine, through which it is to pass in preparation

And here the writer may be allowed to make a remark on the manner of re-working waste, as it seems to him that a great mistake is made just at this point. According to his views of this subject, all the working cotton needs is enough to clean and straighten the fibre and get it in proper shape to spin; more than this is injurious. Now, if this view is correct, what shall we think of taking clean waste, such as is made about breaker and finisher cards, or single cards, lap-heads, drawing, roving and spinning frames, back to the picking-room and mixing it with the new cotton, putting it through all the opening machinery, then through all the cards and other machinery used in the preparation? What is the effect upon the fibre of this cotton, and what is the effect of working this waste in the usual manner upon the whole work with which it is mixed? It must be most demoralizing, rendering it weak and uneven. If any are desirous of knowing exactly how such waste will work when it has had a second run through the opener, picker, breaker and finisher cards, railways and drawing, let such an one try a little alone without being mixed with new, or any other kind of cotton, and he will find out. It will be with great difficulty he will be able to pass it along as far as the last drawing head; and when it has passed through that machine it will utterly refuse to go farther. Take up a piece of drawing two feet long, and it will drop apart; it will not hold together to go through a speeder—*not at all.* It has had too much working, and it must go back to the picker, and a very small quantity be mixed with a large quantity of new cotton, in order to work it.

If this is correct, and no one has disputed it, such waste should not be re-worked with the new cotton. Breaker flyings, and any other dirty waste that needs to be re-worked considerably to clean it, may and ought to be mixed with the cotton when it is picked up. The way to work clean waste will be explained in another place.

When the cotton has been picked up and mixed in the manner described above, it is ready to pass through the first machine. This first machine was formerly in this country, and to some extent in England, a cone willow; such are used now in some places, and they are a very good kind of machine; may be run at about 350 or 375 revolutions per minute, and they are capable of taking care of as much cotton as an ordinary man can stuff through the aperture in the top. They clean the cotton very well, and if the delivery is kept clear, and the waste underneath is not allowed to fill up to the rack, very little injury is done to the staple. But if the bin fills up so as to obstruct the delivery, then they will string the cotton, or as some call it, "roll the cotton." Keep them clear, and they will work very well.

Some use a stick whipper as the first machine for opening cotton. These are excellent machines—cheap, effective, and durable. The speed may be from 175 to 200 revolutions of the driving shaft per minute. One man will whip 12,000 pounds of cotton per day and do it well, too. This machine gives cotton a nice preparation for any machine that is to follow; and it is the opinion of the writer that cotton should always pass through such a machine as this, or some similar machine, before it passes through rigid rolls and beaters. As cotton comes out of the bale in damp flakes it is in no condition for such machinery, and if it is passed through such as the first process, it is done to the injury of both cotton and machinery.

The Calvert willow is still in use in some places, and if cotton is first whipped in a stick whipper, some 700 or 800 pounds may be nicely willowed in one of these machines in a day—that is, if the machine is in its original shape. Much fault has been found with these machines in times past. It has been said that they injure the fibre in many ways; that they make loops, strings, rolls, nits, neps and various kinds of trouble arise from their use. And these charges may all be true to some extent, and yet it may be more the manner of treating the machine than anything else; for instance, forcing too much cotton through them will produce these results. Owing to the peculiar construction of the comb-cylinder, if more cotton is crowded through the rolls than this same comb-cylinder can take care of properly, it will do it improperly. The little grooves in front of each comb being filled up entirely full, more than full, and kept so, the tendency is to roll and string the cotton, instead of loosening out the fibre nicely, as it will do when properly fed.

The Merrimack Manufacturing Company greatly improved these **Calvert** willows **by** building every part stronger, putting them together in such a manner that any cylinder could be taken out without taking the whole machine to pieces, putting the **tooth** racks (or bars) on the wooden cylinders in spiral form instead of straight across, leaving out the third slow tooth-cylinder entirely and raising the front of each comb about a sixty-fourth of an inch, and making the grooves in front of the combs a trifle larger and wider-mouthed. This last-mentioned improvement was the first one made. It was done by **Mr. David Harding**, overseer of the picker-house. First, one cylinder was so arranged as an experiment, and it was found to work so much better that all the others were changed, and it led to the other improvements mentioned. What this Company did others can do if they choose, and they have the benefit of such example and experience. Yet, with all these improvements, the cry was continued that they injured the staple, which was no doubt true in a sense, as these machines clean the cotton better than any other known machine. It is a logical conclusion, that they injure the fibre more than any other. They have within a few **years** past mostly been changed for more modern machines. But the Merrimack Manufacturing Company keep **one to clean** the waste made **by** breaker cards. One of the machines, improved as above described, will clean 2000 pounds per day of ordinary cotton very well; and if the fibre is somewhat injured, it will **do** with less working **in** machines that follow.

In most large concerns at the present time (1875), 'old pickers and willows have been removed, and new and more popular machinery has **been substituted**. The advantages claimed for the new machines (whether English or American) are—

1st. They will do more work. From 2500 to 4000 pounds per day are claimed for most of the new openers; and it is a fact that such quantities are being run through them, and the work is tolerably well done. Some of them are named "Duplex," and others "Compound Openers," because they have double feed—two regular sets of feed rollers and two feed aprons—the grists of the two uniting before they reach the second pair of rollers. The product of these machines is calculated to be finished on a picker, or lapper, with an evener, or self-regulating weight, so that the cotton is spread on the double-feed opener aprons without being weighed, as a general thing. The author has never had personal experience with these machines; but, as has been remarked before, "the principle remains the same"; and as the machines themselves are not materially different in principle from others that have preceded them, they call for the same general management.

2nd. It is claimed that these new openers do the work better than the old ones, or, that they clean the cotton with less injury to the staple, which may be true. It is not for the author of this work to say. It is his business to tell how to manage any machine without particular reference to the merits of the machines themselves. The relation of beaters to rollers should in all cases be determined by the quantity and quality of cotton, and the position they occupy.

The first beaters should be set at a greater distance from the rollers than any of the others that follow, for the reason that the grist is heaviest there, and the cotton has not previously been worked much, and consequently is tough, and works harder there than at any other point. If the feed is very heavy, the first beater should clear the rollers three-eighths of an inch, but may vary from that to five-sixteenths, according to circumstances, and they may be set gradually nearer to each set of rollers down to the last pair in the finisher-picker, the last one not being allowed to run nearer than a tenth or twelfth of an inch.

The speed of beaters should vary in the same manner. They may run at any speed, from 1400 to 1800 revolutions per minute, according to diameter or breadth; the first on lowest speed, and the last on highest, for the same reasons that they are set at different distances from rollers. It is necessary on modern pickers, and an advantage to any kind, to have more or less draught. It may vary from four to ten in different machines. If there are two sets of rollers at any point, as is the case in some machines, it is well to have a little draught between the two sets, say $1\frac{1}{4}$ to $1\frac{1}{2}$. The draught of a picker, as a whole, may be arranged to suit circumstances. The weight of a yard of lap from the finisher-picker may range from 6 to 14 ounces.

Finisher-pickers, or lappers, as they are frequently called nowadays, are calculated to do one-half the work of the modern opener. Two finishers are supplied with laps from one opener. The laps are usually run three into one; on old-fashioned pickers two into one, and in some cases pickers are single.

It is very proper to say that a machine will do so much work in a day, or a week. This means under ordinary circumstances and average kind of cotton. If cotton is very dirty, of course it must be worked more than if it is very clean. The feed of a picker should be changed to suit the quality of cotton, while the beaters and fans may have the same speed for all kinds; and the draught may remain the same. It is better to put cotton through two rapid processes than once slowly; that is, it cleans it better, as much of the dirt in cotton is as light as the cotton itself, though not as tenacious ; it will rattle out if it has a chance, when it could not be knocked or blown out.

It will be readily seen that modern pickers require more draught than old-fashioned ones. It is necessary to have a good fan draught in all kinds of pickers, not only to rid the cotton of dust, but to pack it evenly on the revolving screen cylinders, as it passes through the machine. Without this arrangement a good, smooth lap cannot be made. Most modern pickers have fans attached to their own driving gear, and so stop and start with the machine. Some annoyance has been experienced with this arrangement, where several pickers run in the same room and all discharge their dust into one dust-room. When one stops it takes the dust of others up into the one standing. Mr. Richard Kitson, now of the Kitson Machine Company, has invented a self-adjusting door at the end of the dust-box of each machine, so nicely balanced that when the picker starts it opens, and when it stops the door shuts. Mr. David Harding has fitted a binder to the cone belt which regulates the evener on modern machines, and makes them much more effective than formerly. The last-named improvement has been patented and assigned to Richard Kitson, Esq. The old-fashioned pickers take care of about 800 or 1000 pounds of cotton per day, when run single, and when double (that is, as a breaker and finisher) double that amount. As a general thing double picking is to be preferred to single, for the same reasons that double carding is preferable to single. This subject will be considered in a future chapter, on carding.

A picker, when everything is in perfect order and working all right, makes a round, well-defined hum, like a large bumble bee. If the sound of the beaters is sharp, like the buzz of a smaller insect, it indicates that they are too near the rolls. If, on the other hand, the sound is coarse and fluttering, it shows that they are either too far from the rolls, or running too slow, or being fed too heavily. Of course the sound is not a *sure* guide, but a practised ear will detect a wrong sound very quickly.

And here the author may be allowed to make a suggestion perhaps with regard to the proper preparation of cotton. It has already been said that "all the working cotton needs is enough to clean and straighten the fibre; more than this is injurious." If this is a correct fundamental principle on which to proceed in the preparation of cotton (and the author has never heard it questioned), then there is a chance for improvement in our practice. It is the general custom to take all the waste from the carding from breaker and finisher cards, lap heads, drawing and roving frames, and all kinds of spinning frames, back to the picking mill and mix it with the cotton to be re-worked, and put it through all the other processes of preparation. So far as the waste from breaker-cards is concerned, that method is all right; because that kind of waste is dirty and needs the re-working to clean it. But does the waste made on finisher-cards—such as top-strippings and flyings—waste made about a lap-head, or where a finisher-card breaks down, or waste from a drawing, roving, or spinning frame—all clean and nice

—do these kinds of waste need all this re-working? Most certainly not. It is a positive injury to it, and by such a use of it all the other work is made weak. This is the opinion of the author: it can be taken for what it is worth. He has never tried any other plan, because he has not been situated, for several years past, so that he could. If he had been working independent of all others, and had charge of pickers in connection with his own room, he would long ere this have tried another plan. As it is, his suggestion is to collect the waste from finisher-cards, lap heads, drawing frames, roving frames and spinning frames—all clean—carry it to the whipper and pass it through that machine by itself; then spread it on the apron of a finisher-picker and make a lap, to be used on a finisher-card. Let it be distinctly understood, that this plan is not a revival of the old plan of taking waste generally—dirty and clean—back to the picker and making a lap for the breaker-card, called "a waste card." But this plan is recommended for clean waste, and no other.

An objection may be raised here, on the ground that it is not, or cannot be, well mixed with the other cotton by this plan. But let it be remembered that it had a good mixing the first time it was worked. What need of more mixing? But it may be objected, again, that it would injure the product of the section of cards through which it passed, mingling with the work of only ten or twelve finisher cards, instead of the whole as now. That is true; but would it injure the work of that section, in the manner described, as badly as it injures the whole by the old plan, especially when roving waste is run through a scratching machine and torn all to pieces by the process? This suggestion is thrown in for the consideration of those who may deem it worth their while to think of it.

In the preparation of cotton, like most other kinds of business, it is well to make the most of everything; and to this end it is a good plan, when loose cotton falls down from one story to another —as from the willow-room to the picker-room—to let it pass over a rack, or fall on a rack, or both, as much loose dirt will rattle out in this way and may be gotten rid of very cheaply. It would, perhaps, astonish a man who never saw it tried, or rather, a man who should see it for the first time, to see the amount of dirt that will rattle out of cotton in this manner; and this is a kind of working that does nor injure the staple.

And here, perhaps, a few remarks on this particular subject may be in order. What is meant by the term *injuring* the staple, or fibre, of cotton? Do we understand by it, breaking the staple? That would certainly be an injury; but is that what is meant by it? If it is, there is probably very little if any of that kind of injury done to cotton in modern carding or picking machinery, as the conditions necessary to produce this result do not exist, in proper machinery and ordinary cotton. In order to break the staple, one end must be so firmly held between a pair of rollers as to resist a sufficient power to break it while so held; and our ordinary cotton does not show staple over an inch at the longest, and most of it is shorter than that; while in modern opening and picking machinery the gripping and pounding, or scratching, powers are two inches apart in most cases. When it is understood that the fibres lie in all directions at this stage—as often crosswise as lengthwise of the machine—we must conclude that there is not much staple broken in this way, or in fact in any other way in any ordinary machinery at this date. Probably our cotton is not injured in that manner, viz: by breaking the staple. Most all our common grades of cotton vary considerably in length of staple, or rather, the staple varies considerably in the same lot, even while in the boll, and if we look sharp we shall find nearly all lengths, from one-fourth of an inch to an inch in length in the bale, and at almost every stage in the process of working; but we shall find that the proportion of short grows less as we advance, instead of increasing. Much of it finds its way into waste, in various ways. It is knocked out by the whipper, opener and willow, by the beaters, by card-teeth, and after it gets out into open air, as in drawing, roving and spinning frames, much of it is thrown off by centrifugal force. The air is full of it. After a while it settles on the roller-beams and flyer-plates and on the floor, while the long staple is held in place and goes along in the work.

The author of these pages is of the opinion that we *do not break* the staple, at the present time, if we ever did, in working. But the question, " Do we injure it?" still continues. No doubt every process that cotton passes through, in preparation for spinning, wears it more or less—frets the ends of the fibre, strains and weakens it in a degree, and it is a necessity; there is no getting rid of it, if we clean and straighten the fibre, as we must do if we would have good, smooth yarn.

But in this, as in most other kinds of business, there is a *best* way to do it; and the best way is to work as little as possible and accomplish our object. In the opinion of the author it is *not* the best way to carry clean cotton (waste made on cards, drawing, roving and spinning frames) back to the opening room and put it through all the machines that the raw cotton passes through; but the best way would be to get it into a lap as quickly and easily as possible, put it on a finisher card, and in that manner mix it with the other cotton.

Again: it is not the best way to run roving waste through an extra process beforehand. It is more than probable that a great saving of fibre could be made by a different mode of operation with regard to our clean waste and dirty waste, too, perhaps; but the question of dirty waste will be more fully discussed in the chapter on carding.

One thing more in this connection: The fibre of cotton may be injured by being subjected to the action of beaters and rolls. As it first comes from the bale in damp flakes, it is then too solid to receive such blows; it should be run through a stick-whipper, or something equivalent, to loosen it out a little; then beaters won't hurt it.

The modern picker is a machine which while at work challenges the admiration of the beholder at once. Receiving a lot of loose, disconnected, dirty cotton, and turning out a smooth, even sheet, comparitively clean, and nicely winding it on a roll ready to be handled at the pleasure of the operator, seems almost like magic to those who are unacquainted with the particulars of the plan of operations. The writer was once showing some company through the mills, and while explaining the work of the picker and remarking that in this process the cotton was separated from the dirt, one lady asked the question—"How does this machine know dirt from cotton?" This question of course raised a laugh; but it was, and is, a fair question. How *does* it know? Has the machine intelligence? No: but the builders had; and by employing the elements, and making use of the laws of nature in its construction and operation, the desired results are produced. By a strong fan draught the cotton, as it passes through the weighted rolls towards the beater, is held up against its blades and receives the full force of each succeeding blow; and by these blows from the beater blade the heavy dirt, seeds, etc., are driven down through the rack underneath, or over the "throat piece," as the case may be; but the cotton and light dust are drawn forward by the fan draught; and cotton is packed on the wire cylinders, the dust is drawn through them and finds its way into the dust-room, while the cotton here receives its first shape in form of a lap, and passes on through a succession of rolls, beaters and cylinders until it is wound up in convenient form to be placed on other pickers, to pass through the same process again, and come out in a condition to be placed on a card. The credit of this machine, in its present form, is not due any one man, or two, or three, but has come into being through a succession of efforts by different men for many years. One man has conceived something in his own brain that would improve it in one particular, at one time; another, something else at another time; and yet there is room.

Like most other machines there is still a chance for improvement; and there are men now at work in this direction, with as much zeal as any who have preceded them in the same kind of business, and their labors are being crowned with as much success. Among the indefatigable laborers in this direction may be mentioned Richard Kitson, Esq., who has been engaged in the business for the last twenty years, and has brought out a machine, or rather a set of machines (openers and lappers), equal to any of this class in the country; and still he is at work on improvements. Mr. Kitson formerly did business on his individual account, but it increased to such an extent that it became necessary to form a company, and such a company was formed in July, 1875, with a large capital. Mr. Kitson is president, and S. E. Stott treasurer and agent. The company are now at work on an improved opener. It differs somewhat from their deservedly popular compound opener. Instead of the double feed (two aprons and two sets of rolls and beaters) the cotton is taken from the bale, and after being mixed, is placed on two narrow aprons, each half the width of the opener frame, and run side by side with a partition between. When it arrives at the frame, it drops a little over the end of the aprons, then turns a sharp corner, half to the right and the other half to the left, and is facilitated in this movement by a little rough roller and a strong fan draught, up to an arch-like entrance into what appears like entering the end of a cylindrical-shaped box above the centre. Inside this same cylindrical-shaped box are two wrought-iron circular plates, running side by side, and close together, and in opposite directions. One is a little smaller than the other. These plates are armed with wrought-iron bars firmly riveted in a circle not far from the edge of each, so that the arms of the larger plate project over the edge of the smaller one; the smaller plate has a similar circle of bars or arms, the two forming circles of projecting bars, one inside the other, a few inches apart, and running in opposite directions, all projecting towards the entrance where they receive the cotton, which is held up to their ends by a strong fan draught until the fibre is somewhat loosened out; and as it becomes loose it enters the box, receives considerable of a stirring up between these double sets of bars, and is passed along to the beater in the usual manner.

This new arrangement is called an Atmospheric Opener. The speed is about 700 revolutions per minute.

The first beater in this machine is also a new invention. It is a three-knife beater, but instead of the blades running straight through, from one end to the other, in the ordinary manner, they are put on in three sections. The whole thing perfectly balanced and dealing continuous blows, runs very steadily, with less jar than is usually made by beaters. Another new thing about this beater is that the arms, instead of being drilled through the shaft, as is common, are so arranged by means of collars and packing boxes, that they will yield a little when they come in contact with anything unusually thick or solid. This is called a Sectional Elastic Beater. It runs about 1200 revolutions per minute. The Company have one of these new openers at work in the Boott Cotton Mills. It is claimed by the inventors that this opener will do its work in a satisfactory manner, with less severity of action on the fibres of cotton than any other modern opener. The action is similar to that of the old "stick whipper." It certainly commends itself to the judgment of the writer, as it is his opinion that cotton from the bale should not be subjected to the action of rolls and beaters as the first process. This new opener has come to the knowledge of the writer since he commenced this chapter; and the new invention, together with the fact that he has for the last two or three years used laps made on the Kitson picker (the Merrimack Manufacturing Company, at Lowell, have adopted the Kitson machines within that time, excluding all other lappers), must be his apology for this digression. A good, even, clean, well-wound lap is a very essential starting-point in carding; and that machine, or set of machines, that will produce this result with the least wear and tear of fibre, is the most desirable.

With all opening and picking machinery, it is necessary to set everything properly and firmly, and to keep machines in order. They need the constant, watchful care of a living man, and he should be wide awake and active.

There is some danger of fire in a picker-house, and as the material is so very combustible, it is necessary to be constantly on the watch. Most picker-houses and carding-rooms are amply provided with facilities for extinguishing fire, and certainly all should be. Many have sprinkler pipes running all over the building, that may instantly be filled with water and produce a shower all over the room. It has sometimes occurred to the writer that there should be some unmistakable mark by which any one at a glance may see whether the sprinkler gate is open or shut, as it sometimes happens in case of fire that one man opens the sprinkler gate, and immediately another man, not knowing that any one has preceded him, thinks he opens it when he shuts it instead. Another discovers that the gate is not open, and so he proceeds to open it; and by the time the water begins to flow through it the second time, the pipe has become red hot, and is broken as the cold water strikes it, and the use of the sprinkler is lost for that time, when if there had been some plan of the above description the property might have been saved.

In case of fire in the picking or carding department, what is most needed is very prompt action. Suppose a spark of fire drops from a bearing overhead on to the floor among loose fibres of cotton. Some one sees it. If he stops to get a pail of water, or to go after a blanket some distance away, by the time he gets back there will be likely to be more fire than he can manage, either with a blanket or a pail of water. The most effective means of extinguishing a little fire, when it first strikes cotton, is to smother it; grab it right up in the hand quick as thought, then it may be carried to a pail of water at leisure and submerged. If fire is discovered inside a frame or box, or any other enclosed place, don't open the doors until you have something in readiness to extinguish it with. Get the water and the blankets ready first, in that case. Fire will not burn much without air, and a little air sometimes helps to spread it most wonderfully.

Gas pipes should be brushed off clean every evening before lighting up, as the loose fibres of cotton that are continually flying around the room sometimes gather on gas burners and pipes in sufficient quantities to make a fire, if a blaze comes in contact with them.

CHAPTER III.

Carding.

GENERAL REMARKS — CLOTHING CARDS — POWER REQUIRED FOR FILLETS — HOW TO ASCERTAIN IT — TOPS — DIFFERENT KINDS OF CLOTHING — GRINDING: OLD STYLE — IMPROVEMENTS: HOW MADE — PARKER GRINDER — FURTHER IMPROVEMENTS SUGGESTED — GRINDERS: HOW TO MAKE TRUE AND STRAIGHT — HOW TO CLOTHE A GRINDER — GRINDING CARDS — CARD CLOTHING RISING — HOW THEY ACT — HOW TO CURE — SETTING UP CARDS — THE QUANTITY — DOUBLE OR SINGLE CARDING — DISTRIBUTING DRAWING IN RAILWAYS — STRIPPING CARDS — SCREENS — SPEED OF MAIN CYLINDERS — LICKER-INS — DRAUGHT OF CARDS — GRINDING — CLOTHING GRINDERS — OLD EMERY — RE-CLOTHING CARDS — PATCHING UP — WASHING CARDS — IMPROVEMENTS IN CARDS — A NEW GUIDE — WORKING WASTE.

However nicely cotton may be opened and prepared in laps, it needs carding; of course the nicer the preparation in the shape of opening and picking, the less carding it needs; but it must have some. Picking is a kind of wholesale business, as compared with carding. It has in this process been handled by the lump; it must now be handled more in detail; individual fibres must be attended to, and nothing but properly adjusted card teeth will do this. As this process is *one* of the most important, if not *the* most important, process cotton passes through in its preparation, it will be dwelt on in all its particulars. Remarks on machinery in the first chapter of this work should here be borne in mind.

Before any card cylinders are clothed, they should each be covered with good, stout, even cotton cloth, put on perfectly smooth, either with thin glue or size. After they have had time to dry, the little black specks and other nubs should either be shaved off with a sharp knife, or hammered down smooth; then the cylinder is ready for the clothing. A breaker main cylinder should be clothed with clothing stuck with No. 32 wire, in well-tanned and nicely prepared leather, A No. 1 in quality. There is very little danger of drawing sheets on a main cylinder too tight. When they are new, of course such a thing is not impossible; but where this is the case once, it will probably be the other way a hundred times. Nothing is more discouraging to a carder, or injurious to owners, than to have card clothing " huff," or rise, in a few months after it is put on, and necessitate "drawing over." It is a waste of time, a disagreeable job, loss of the work of cards, expensive, as well as keeping the grinders from their legitimate business, and it is destructive to clothing. After a main cylinder has been drawn over its death warrant is signed; the next thing that may be looked for is breaking out, and a kind of breaking out that continues until it becomes necessary to remove the clothing and put on new. As a general thing sheets should be drawn when new as hard as the leather will bear. Of course drawing over is sometimes necessary, and when it is so it must be attended to, and then be careful and *not* draw too tight. Remember, the stretch is out of the leather now, and it must be dealt with gently or immediate breaking out may be expected. *Draw easy the second time.*

Fillets, with which doffers, licker-ins, and sometimes main cylinders, are clothed, work very differently from sheets They are continually working loose. When they are at work they go one way, and while grinding they go the other. It is possible to draw a doffer on too tight; and if they are so drawn they will be very likely to break out more or less while grinding. It is well to draw a doffer twice in clothing; that is, nail one end, and wind the fillet on loosely; then wind it back on the drum; then apply weight enough to the friction strap to make about forty-eight pounds' power or strain on the fillet, and in this manner carefully wind it on the doffer cylinder; then reverse and wind back on the drum, keeping it tort all the while. Before winding on the last time, reduce the power to about thirty-six pounds. The first winding with power takes the stretch out of the leather, and the second with thirty-six pounds' power will be tight enough to grind well; and it has been found by the author that doffer fillets put on in the manner described will run longer without becoming loose than they will put on in any other way. When a doffer runs loose, as they are likely to do, it will generally be detected at the end of the cylinder while being ground. Two or three coils may be drawn over by hand and make the whole thing all right, as the remainder will frequently be tight enough while the end may be loose. But if there is a general appearance of being loose all over—edges huffing, with an inclination to rise, either while at work or grinding—then take it out of the card and put it in the machine and draw it all over, applying about twenty-eight pounds of power; not more for an old doffer.

The power, or resistance, of a drum may be ascertained in the following manner: Put the friction strap over the pulley and weight it; then wind a few coils of narrow belting, or a small cord, round the drum, confined to it at one end, and let the loose end hang over in the direction that the fillet will draw; then, by means of hooks, attach weight to it—28, 36, or 48 as desired, and remove the weight from the friction strap until the drum moves, allowing the same weight to remain on the friction strap while winding the doffer fillet from the drum to the cylinder; and this weighing of power must be repeated often, as the same weight and strap are not sure to give the same power on two different days, owing to a different state of the atmosphere. Licker-ins, if the fillets are set with diamond wire and need no grinding, may be drawn as tight as the leather will bear. The teeth seldom break out. If they are set with round wire, and consequently must be ground, the same plan recommended for doffers may be pursued with them.

Tops, before they are clothed, should be carefuly jointed, and the plates set out of wind; and when old clothing is removed, before new is put on, they should go through the same process, as wood cannot be depended on to stay exactly where it is left year after year. Top clothing needs very little drawing. No ratchet, or treadle, is necessary. A weight of from 6 to 9 pounds, according to circumstances, may be attached to the strap running through the levers of the clothing pliers, and is sufficient, convenient and gives the clothing a uniform drawing. If there is thin, or spongy, leather, it will of course draw down a little more; but never mind; it is of more importance to have the clothing properly drawn than it is to have a nice, straight edge simply to look at; and the same is true of all other sheet clothing, whether it be main cylinders, tops, or strip cards; and strip cards need less drawing than tops, because the leather is thinner and the teeth not so closely set, and they do the work of stripping better to be a little loose. Every class of cards, and every card, should be carefully set up after the clothing is finished, whether it is to be ground or not, as teeth that are out of place are rather a damage than a benefit to any card.

The clothing for tops on a breaker should be set with different sized wire, according to position.—Nos. 1 and 2, No. 29 wire; Nos. 3 and 4, No. 30 wire; Nos. 5 and 6 with No. 31 wire, and the remainder with No. 32 wire—same as main cylinder and doffer. A licker-in should be set with No. 26 diamond wire. When cards are clothed with fillets, a scraper should be run round between the coils, to make sure that there is no lapping of one coil on the edge of another.

All the above has reference to leather clothing; there are many other materials used to set card teeth besides leather; but as these vary so much—some very good, and others very poor; some giving no trouble whatever, and others always troublesome—no general rules can be given; the carder must be governed by circumstances.

GRINDING.

After clothing comes grinding; and there are about as many different opinions as to the best method of grinding as there are different men who have this business in charge. That this branch has been greatly improved, within the last thirty years, no one who has had experience in it will deny; and the end is not yet. Many of our improvements were discovered by accident, or were brought about by the force of circumstances. Under the old system, when the driving pulley on the main cylinder was 24 inches in diameter, the driven pulley on the grinder 2 inches, and the speed of the main cylinder 120 revolutions per minute (making the speed of the fancy grinder 1440 revolutions in the same space of time): those were days that tried men's patience and sadly injured cards. It used to require from four to six weeks to grind a new card; then two or three days must be spent in "strickling," to take the barb off the points of the teeth, and it was poorly done at that; for when cards were new, it was expected that they would have a "wire edge," as it used to be called. They did not card well, strip well, or do anything well. When cards of a more modern plan began to be introduced, with driving pulley on the main cylinder 16 inches instead of 24, we used to grind better, quicker, and our cards worked better, stripped better and were better, in every way. The speed of the grinder was reduced by this arrangement from 1440 to 960 revolutions per minute.

But was this change made for that purpose? Probably it was not; but it did benefit us in that particular. Probably few carders ever attributed it to slower speed. The author of this work did not. When the Parker traverse grinder was introduced it was found, by actual trial, that one of this kind (7 inches in length, instead of 37 inches —the length of the fancy roller grinder) would grind a card much quicker and in a more satisfactory manner than the fancy grinder ever had done the same work. Carders began to inquire into the reason for this. Something was claimed for the new kind of traverse, and something belonged there, no doubt; but it was evident that the new traverse should not be credited with the whole improvement. There was something else to be taken into account; and that something else was the comparative low speed of the Parker grinder. It had a 5 inch driven pulley instead of a 2 inch, making the speed 576 revolutions per minute on the 24 inch driver and 384 with a 16 inch driver. Probably Mr. Parker never thought that slow speed would facilitate grinding; but, on the other hand, it is very likely he made calculation to drive his grinder as fast as it could be driven, consistent with its construction. Happy for him and all manufacturers, it would not admit of the old-fashioned high speed; and to this fact alone much of its superiority over the old, long grinder may be justly attributed. This is not all: We have learned to drive all kinds of grinders slower than formerly, and we are constantly reaping the advantages which this system affords. And have we, at this date, arrived at perfection in this particular? Who shall say we have? Perhaps from 350 to 400 revolutions per minute is slow enough to drive a 5 inch grinder; but it has been the opinion of the author, for some years past, that we might make another equally valuable improvement by driving our cylinders slower and allowing the grinder to retain its present speed. In nearly all other kinds of grinding and polishing, with either emery or stone, the emery roll, or stone, is driven at high speed, and the article to be ground, or polished, is driven slow. How is it with grinding spindles, or other iron and steel rods, or cylinders, with a stone or emery? The grinder moves fast, and the article to be ground moves slow. Why should grinding cards be an exception to all other kinds of grinding? Can any one tell?

Several years ago the author experimented considerably in this particular. In 1866 he had an extra stud fastened in a large plate and the plate fixed to the frame of a card about to be ground, and on the stud were two sets of pulleys—one set driven from the main shaft overhead, and the other set smaller, which drove the main cylinder, thus reducing the speed of cylinder while grinding; and the result was that cards ground by the slower process were ground better, worked better and gave better satisfaction in every way. He is still experimenting; he has not yet arrived at a point where he can say, "Eureka," but he is satisfied that a slower speed, at some point below where we now grind cylinders, would be beneficial.

A small counter-shaft might be placed immediately over the card, from which the cylinder, doffer and grinder could be driven at any desired speed. The present pattern of the Hardy grinder, with the small pulley on the screw, necessitating a tight belt, often causes a little jump in the grinder when the lacing, or hooks, pass over it. The remedy is a large pulley, both driving and driven. Care should be taken, in adjusting any kind of a grinder to a card, to have all the nuts and bolts thoroughly screwed up, so that the grinder may remain firm in place while grinding. If it gets loose, it makes very bad work in a short time.

The first step towards good grinding is a good grinder, properly made, perfectly true and straight, and perfectly balanced. If it lacks any of these qualities it will not be a good grinder. Grinders constructed of tin are seldom quite round. To make them so, first paint and let it dry. Prepare a leather fillet three fourths of an inch wide and fix it on the grinder with glue, giving the fillet all the strain it will bear. When it has had time to dry, place in a frame and run it in its own bearings or boxes, and drive it with its own pulley at a high speed—as high as it is calculated to run while grinding. Fix a rest its entire length; then with a sharp chisel, or plain iron regulated by a guard outside the rest, turn it off round and straight; paint the leather, let it dry, and repeat until the grain is filled and the surface smooth. When this is accomplished, put on a coat of old copal varnish and wind it with twine, while the varnish is green. After it is dry it is ready for the emery. A grinder prepared in this manner will last years, if care is taken with it. Especially be careful not to let it lie long in hot water, in removing a coat of emery.

All kinds of grinders should be painted and wound with twine previous to clothing with emery, whether they need any leather or not; and the twine should be wound into green varnish in all cases. The best emery for clothing grinders is No. 10 English emery; and only one coat should be put on at any one covering. No sizing should be used on a coat of emery, if it is desired to have it grind well. If it is sized, it will soon glaze over and become an old, worthless grinder, while it should and would be a good, effective one without any size.

It is sometimes said by carders that their "emery won't stick" without sizing. [See recipe for making glue in Chapter Sixth.]

In covering a grinder everything should be in readiness beforehand, so that there may be no delay after the work of clothing has commenced, and it is better to have help enough to rush it a little. Spread the glue evenly over the surface of the grinder and put the emery on copiously, keeping the grinder moving slowly all the time. After the emery is all on, roll it with an iron roller and keep the grinder moving slowly round for a half-hour or more after the clothing is all over. This prevents the glue from settling down on the lower side of the grinder, as it will surely do if left at rest immediately after the emery is put on. A grinder should be allowed to dry 48 hours after being covered before using. When first started hold a piece of pine wood on it, to knock off any high kernels of emery that may adhere to it.

It is important in grinding cards to have good men to attend to it, as well as good grinders and grinding machines; both are called grinders—machines and men. A man in order to be a good grinder must have good eyes and quick ears, especially the latter, as much depends on hearing in adjusting a grinder properly. He must be a man of a mechanical turn of mind, and must be interested in his work. The best grinders are those who commence to work about cards in youth and are promoted from one position to another until they are promoted to grinders. In this manner they may, and do, acquire a sort of general knowledge of the business before actually entering upon it that is almost indispensable. A new hand—a man who comes from outside, from other business—will rarely make as good a grinder as one who has been brought up in the carding-room.

The time required to grind a new card, or an old one, depends upon circumstances. If the cylinders are perfectly true and in balance, the grinder good, properly applied and the clothing even, it will grind very quick—sometimes in 10 or 15 hours. If any of these good qualities are lacking, it will take longer, and in proportion to the deficiency. When the cylinders are out of true, or out of balance, as is often the case, it will and often does take from five to twenty times as long to grind a new card as it would if all were right. And that is not the worst feature of it: in such a case one side of the cylinder is often necessarily ground more the first time than would be necessary to grind it in ten years if it had been true and balanced. Good, nice cards are injured and often break out the first time grinding for the above reason, and no other. But they must be ground until all the parts are finished.

There are other causes of cards breaking out besides being out of true, or out of balance. The wire of which the clothing is made may be poor;

there may be a hard splice in the leather; they may be ground too hard or too long, as they often are. Some of these causes the carder has no control over, and knows nothing of until the results appear; and when they appear, they must be met as best they can. But a carder should insist on having his cylinders true and balanced, on having good clothing, and grind no longer or harder than is necessary.

Sometimes card clothing rises and causes great vexation. There used to be an opinion that there was something mysterious about this matter—something that could not be explained or even understood. Rising is not confined to any particular kind of clothing or cylinders. Sheets, fillets, main cylinders and doffers, all rise, or are liable to rise, and one as much as another. But different kinds require different treatment. When sheets rise, they must be drawn over; there is no other cure. It is sometimes curious to see a little patch rise in a sheet when all the rest of the teeth are down in their places—perhaps two or three little patches; it may be half-a-dozen, more or less—in some instances only a few teeth in a place. When this is the case, the high teeth are forced back with a card-brush, a gauge, and sometimes with a piece of soft pine board; and then the cylinder is turned by hand or started very carefully, until filled with cotton; afterwards they will sometimes run until another grinding. At other times, this has to be repeated until the patience of the grinder is exhausted, and he calls his overseer or the second hand; and the two, and sometimes three, work for hours together to get a card of this kind started. There may be a little refractory group of teeth that will rise every time the cylinder is started until the operators are all worked up into a fever, when perhaps an excited carder takes a hammer and pounds them down solid and walks away to his bench, declaring there is something mysterious about card-teeth rising; that he never saw any one who could explain it; he don't believe that anybody knows or ever will know why it is.

A main cylinder clothed with a fillet will sometimes act about as bad, and the same is sometimes true of a doffer. The author of this work has had experience in all these kinds of rising; and although he has sometimes been terribly tried, he never pounded down any card-teeth with a hammer; but he has set himself at work to find the cause, and after a long and diligent search found it, and is now able to cure every case with comparative ease. But he prefers preventive to cure. If main cylinders clothed with sheets are drawn tight enough, when first put on, they will seldom rise; if they do, they must be drawn over, as teeth are liable to rise if the leather is loose; *but they can never rise if it is tight.* The author once clothed some cards in a hurry with clothing made in a hurry, and poor stock at that, as it proved. These cards (especially main cylinders) after being ground round about three times began to rise in the most annoying manner. The grinder was directed to remove the tacks from the back of a sheet that was in trouble of this kind, and do it so carefully as not to disturb the teeth, and allow a peep to be taken at the underside, to ascertain exactly how they looked and what position they were in. This peep explained the whole mystery. First, the leather was loose and had risen up by centrifugal force; second, some of the teeth went up with it, retaining their position; third, the leather had slipped up on other teeth and left them, the lower ends resting on the cylinder;

fourth, the teeth that **were set tight** enough in the leather to retain their original position were the ones that we supposed had **risen,** and those that lay on the cylinder seemed as they usually appear when they are all right.

It will be readily understood that drawing over is the only remedy; but it should be carefully done; there is danger of drawing cards too tight the second time. If the teeth all retain their places in the leather, then **it** is said that the sheets huff up, and there is no mistake about it. Such readily show when they need drawing over. There is another trouble with sheets: sometimes there will be a spot in the middle huffed up when all the **rèst** of it seems to be right; and it is frequently the case that such sheets are drawn over and the trouble continues. Such cases are usually caused by a soft or spongy spot in the leather, and the rest all firm; in such cases the drawing over does not affect it. When this occurs search out the boundary line of the soft spot in the leather; draw out a row of teeth on either side; slit it up with a knife; then draw that particular place, and the trouble is cured.

Fillets will rise occasionally when they seem to be tight enough, and are tight enough, and the cause of the rising is, they crowd. Cloth fillets are more apt to do so than leather, but the latter is not entirely free from it. A strip of cloth or leather, either by frequent pulling or straining, gets loose on the edges, and when **two** edges come together there **is a** tendency to rise; and these are often drawn over without removing the evil. When a fillet acts in that manner, draw over lightly and lay **it apart**; let a **thin** piece of iron **run** between the edges as it **is** being drawn, and the trouble will disappear.

Cards having been once ground down need but little grinding at any one time afterwards, unless they get jammed, faced, have to be drawn over or something unusual happens to them. The old plan used to be to grind about so long any way. Some used to grind two days, others a day and a half, and others still a day. This plan is all wrong. The more a card is ground, after it has been brought to the right condition (that is, the teeth ground to a point) the worse. They may need to be ground an hour or two, perhaps three hours, but never grind a minute after the teeth have been brought to a sharp point. It is difficult to grind too often, but very easy to grind too long. Cylinders do not usually suffer so much from over-grinding as do tops, as they are ground on a more scientific plan. The cylinder moves quickly and the card moves slowly over it, and as a consequence they grind faster, and before the grinder is aware of it they are often—yes, almost always—ground too much. The writer has seen tops spoiled in a few months— ground down to the knee. Grinders should be constantly cautioned about grinding tops. Sliding across the grinding cylinder about a half-dozen times is sufficient, as a general rule. It is well to have two grinding machines set close together, where it is practicable to do so, and require each grinder to run two while grinding a set of tops. They will usually grind, in this manner, about as fast as they can be changed. Licker-ins, if set with diamond wire, need no grinding, unless they get badly faced.

In setting a card up ready for work, first clean it nicely—then set the doffer as near as it will run without touching the main cylinder and secure it firmly, so that it will not get out of place while at work. Next set the tops, commencing with those nearest the doffer. The proper way to set tops is, to set the front twice the distance from the cylinder that the back is set; this plan brings the nearest point about two-thirds from front to back, and this part should be set as close as it will run, down to No. 5, and from there to No. 1 set off a very little farther; in other words, tops should be set so that they will fill even. If it is found that they load too heavily on the front, raise them a little; and on the other hand, if the fronts fill scant, lower them. Care should be taken to have every top set square on all the screws, so that there may be no rocking. If workers and strippers are used, they should be set as near as they will run without touching, and the same of licker-ins. On a finisher-card the top feed roller should be set as close to the cylinder as it will run, and no play should be allowed to the caps in the roll stands. On breakers, where licker-ins are used, set the bottom roll the nearest—the same as a top roll in a finisher.

When everything has been adjusted ready to start, turn the cylinder backwards by hand to make sure that it don't come in contact with anything about it. This precaution will save many a card being faced, as it takes but a short time when a cylinder is running at full speed (points first) and rubs the doffer, tops or rollers, to spoil the points, and the work of grinding has been all lost in that case; and worse than that, it has sustained a permanent injury. Set combs just near enough to clear the doffer.

THE AMOUNT OF WORK FOR A CARD.

As to the quantity of cotton a card will work in a day, there are, perhaps, more opinions than about any other one question. We find them carding from 30 to 160 pounds, under different circumstances and conditions according to the kind of card used and the kind of work to be made. There is also a great difference in the amount carded with the same machines, and the same kind of work in different places and under different management. The question whether it is best to card cotton once or twice is still unsettled and open for discussion. There are many very strong advocates of the former plan, and perhaps just as many and just as strong advocates of the latter plan. Many manufacturing establishments have changed from single to double carding, and perhaps an equal number have changed from double to single; and still others which have changed back and forth more than once, while some have both kinds at work. With all these trials cotton manufacturers as a class are undecided which plan is best. There are individuals, and some manufacturing concerns, who think they have settled the question so far as they are concerned; and no doubt they have. But as some have settled it one way, and others another, the inquirer is still in the fog, because each party who thinks he has settled the question claims that his way is right and that the others must be wrong, so the discussion goes on.

Perhaps a little reflection on the part of those interested would go far towards solving this problem. If the position assumed in this work is correct, viz: that cotton needs to be worked just enough to clean and straighten it, and no more, the question naturally arises at this point—Are we carding cotton too much now? If we are, it is a very easy matter to card less, or rather, put more cotton through a card than we are now doing; and this can be carried to any desired extent without resorting to single carding. On the other hand, if we are not carding cotton enough to accomplish the desired result, we can put a less quantity through each card, though of course this latter plan would call for more cards.

Are we carding cotton too much? Let the cloth answer. Are there not black specks enough in it? If we find by examination that there are not, we can increase them by carding heavier—giving each card more work to do—and it matters but little whether it is single or double carding, but the little difference that it does make is in favour of double carding. By the double process we get a better mixing, which is certainly an advantage if that was all. But it is not; for by the double carding process we reverse the fibre and operate on both ends instead of one; and that is a second advantage over the single process. The author is aware that some of our English friends advocate as little reversing of staple as possible in the process of preparation; but he is not yet convinced that this theory is correct—indeed he is strongly of the opposite opinion, and has arrived at this conclusion after a fair trial of both plans. Moreover, he is of the opinion that whoever will give this matter a fair and impartial trial will come to the same conclusion.

The contrast between double and single carding is best seen in the yarn and cloth; and let it be understood that this difference is always noticeable to a greater or less degree where everything is managed in a proper manner—single cards doing half the work of double ones each, as they always should do, other things being equal, wherever and whenever this system is adopted. What has killed single carding in many places has been trying to do as much on each card in this manner as double ones are accustomed to do. It will be very readily seen that by this plan the carding is increased 50 per cent., which is rather more than it will bear profitably, as a general thing. As has been before remarked, if we are carding too much we can card less. But where is the concern that can stand an increase of 50 per cent. and live through it? Many have tried it and signally failed.

The difference between double and single carding is: First, rougher yarn from single than double; second, dirtier yarn; third, spinning does not run as well; fourth, the cloth is not as smooth, or clean, from single as from double carding. It is true that yarn made from single carding, other things being equal, will stand a greater tension on a yarn-tester than double; and the reason may be found in the fact that the fibre has not been worked so much. But this advantage is more than balanced by the amount of dirt in it, causing it to run badly through every process it passes. Another advantage of double over single carding is, the cotton has to be handled over more times, and in every handling some dirt is rattled out, and without damage to the fibre either. Of course, in estimating the amount a card will do in a day or a week much depends on the kind of cotton used. Dirty, nubby cotton requires more carding than clean. Much depends on the kind of card used; the greater the working surface and the more cylinders, tops, and workers, the more it will card. There are a great variety of cards and likely to be more before there are less. Two gentlemen of this city (Lowell, Mass.) have invented, and are now (October, 1875) building a new kind of card; the cylinder 36 inches; but it has two licker-ins and double the number of tops in a Wellman self-stripper. It remains to be seen what it will do, but it seems to be a step in the right direction. The name of this firm is Foss & Pevey.

The kind of goods to be made has something to do with the amount of cotton carded in a given time. If the goods are fine and light, it is necessary to card more than if heavy and coarse. On print goods 64 x 64, seven yards to the pound, a 36-inch card will do a pound and a half to an inch in length of cylinder per day on ordinary kind of stock, and do it very well, too, double carding. If single, then half the above quantity. When cards are kept sharp and in shape every way, they will do very much more work (and do it well) than they will if allowed to run dull and out of place. The beauty of a card of any kind is to have it so arranged as to readily seize, and as readily let go, the cotton, and if they do the first, they will usually do the second; and to do either, there must be a fine needle-point on the teeth. If the points are barbed, or as denominated, have a wire edge, they will do neither. Cotton may be forced through such cards, but it will not be carded. It will be ground or jammed through, and it is impossible to make good work of such.

A licker-in is a very useful appendage to any card. The stout, diamond-pointed wire, or the saw teeth, such as are in use in some places, take hold of the cotton first, loosen it out and throw considerable dirt down on the floor. It is well to cut out a space, five-eighths of an inch wide, under the bottom feed roller, to allow the dirt to fall outside the card-frame, instead of inside, to be mixed with the flyings, making it necessary to pass them through other machinery to again separate them from dirt. When it has once been thrown out (in the manner described) we have done with it, and it will be swept up and carried to the waste-house, where it belongs, and the licker-in is no longer troubled with it, but can, and does, pass the cotton to the cylinder minus this dirt, thus saving wear and tear of card clothing, and leaving it at liberty to do its best with the cotton. The above idea originated with Asa B. Lyford, a carder on the Merrimack Corporation, in Lowell.

Goodwin & Atkinson's Patent Mote-Collector is a useful appendage to a finisher-card. It is much more effective on finishers than breakers; or, in other words, it is more effective where there are no licker-ins than with cards that have licker-ins. A strip of common cotton cloth, four double and about an inch and a fourth wide, so fastened to a self-stripping card as to just touch the top, will keep them very free from dust and lint, which it will collect in a roll that may be removed twice a day. It also prevents much dirt from being brushed off by the hands who clean cards, which is usually brushed or blown into the gearing, and into the work again.

Great annoyance has been experienced by carders in not being able to properly distribute the drawing on the railway aprons in both breakers and finishers; and it is quite as important in the one as the other; for if the cotton goes up to the lap-head uneven, it will make an uneven lap for the finisher to work, straining the clothing with the thick places and leaving the main cylinder and doffer streaked, and making bad work generally, while in the finisher-boxes, if it is not evenly distributed, it tries the top rollers in the railway head and will not draw evenly, and will soon injure the rollers so that they will not do good work. Kent's Card Guide is a very great help in this particular. But the humble author of this work invented a card guide, some four or five years ago, that is better than any he has ever seen. There is no patent on it, and probably never will be, and there is only one at work at present, and there never have been any more; but if the world stands a while longer, and cotton manufacturing continues, there will be likely to be more of them in use. A gentleman made some changes in the author's guide, and there are a number of those at work in the Merrimack Manufacturing Company's carding-rooms, which are working very well. **But** the original, as built by the author, possesses three important advantages over all others that he has seen. First, it is less expensive than any other; second, it **is** more convenient; third, it is more effective, doing exactly what is desired to a hair's breadth, while the others have to be moved just so much, if moved at all. It is claimed by some carders that if the ends are once properly distributed and put in right shape they will always be right; but a little reflection will convince any one that we need an adjustable card guide—one that by the touch of the thumb can be changed—as almost every day a different card in the same row is stopped to grind, causing a derangement of the whole sheet unless some convenient plan for re-adjustment is adopted. The card guide spoken of can be seen at work in No. 6 Carding-Room, Merrimack Manufacturing Company, Lowell.

Another source of annoyance to carders, and waste to owners, is the cotton dropping down at the ends of the doffer on to the railway box or the floor, as the case may be. This trouble can be remedied by a little care in clothing **new** cards, or a little change in old ones. If the clothing on the doffer is a trifle shorter than that of the main cylinder, the trouble ceases. Where the clothing of the doffer reaches out beyond that of **the** main cylinder, **the dropping takes** place **as the** currents of air produced by **the** revolutions **of** the main cylinder blow some cotton on to **the** surplus ends of doffer clothing, and it is knocked off by the comb on to the railway box or floor, instead of going into the work properly carded as it would, and will go, if the two surfaces agree, or if the doffer is slightly shorter than the main cylinder.

STRIPPING CARDS.

Stripping cards has undergone considerable change within the last ten or fifteen years in more than one particular. The old style used **to** be to strip tops by hand once in 12 or 15 minutes—half the tops on each card. **It is** now mainly done by power. Wellman's and other self-strippers have come into very general use, and very properly, too, **as the** work is done much cheaper than formerly **and** quite as well, and the speed can be regulated so as to strip more or less often to suit the taste of those **who** have the management of them. The old time has been doubled, or rather the stripping has been doubled **in** most places. One-half of the tops **are** now stripped by power once in 8 to **10** minutes. The author is of the opinion that one-half the present stripping might profitably be dispensed with. It may seem to be a strange idea to many, **but a** little examination into the facts of the case will certainly do us no harm, and we may possibly learn a valuable practical lesson from such an examination. Let any one who thinks it worth his while try the following experiment: Take a top from any part of a card, front or back—a top that has run its full time and ought **to** be stripped, according to rule, because it has been working 10, 15 or 20 minutes, as the case may be; strip it clean and carefully with a hand card, weigh the waste stripped off on a nice pair of roving scales and write down the number of grains it weighs; then replace the top and let it remain in place exactly one minute by the watch; remove and strip as before, weighing the waste, and it will be found that it weighs 50 per cent. as much as the former weighing did. Now, if this is the case (and no one is asked to believe it until he has tried it), what does it show? Simply that the clean top helped itself to the first cotton that came along by it until it was loaded to a certain degree; and if the waste is examined, it will be found to be almost as good cotton as there is in the room. Suppose we strip all the tops as rapidly as possible by hand and continue the operation a few minutes, we shall find we have greatly lessened the weight of the drawing delivered by the card, whether it be a breaker or a finisher. What does this show? That a great part of the cotton can be stripped off from the tops, if they are stripped often enough. Well, what have we learned by these experiments? We have learned this, it would seem: that it takes a certain amount of cotton to fill a top up to a working condition, and that certain amount of cotton is as good as any that is being carded; and, also, that the top will continue to catch dirt, nubs, leaf, and nits, just as well for a considerable time afterwards. Who knows how long? Evidently more than 36 or 46 minutes. This is probably a new idea to many if not to most of carders and manufacturers in general. But it is worth trying, and may prove useful. If this theory is correct, it is no use talking about stripping the lower tops oftener than those higher up, as has been advocated by some. The author has allowed cards to run two hours with the stripper standing, before perceiving any difference in the working of the card. **He** will not lay down any rule to strip tops by, but thinks the time now occupied may profitably and safely be double the present, or in other words, strip tops one-half as much as they are stripped now—profitably, because good cotton will be saved, the percentage of waste will be smaller, and the wear and tear of machinery less; safely, because he thinks they will remove as much dirt as they do now.

There is one serious objection to putting the speed of a Wellman stripper down very low, to secure slow or less stripping. If it is slow it keeps the top off the card too long. If some plan could be devised to move the stripper slowly from one top to another, at the same time having a change of speed to do the stripping, so as to rise, strip and replace the top quickly, a great advantage would be gained. Perhaps there are such strippers in the country; there will probably be more. Mr. George H. Chandler, assistant superintendent at the Merrimack Manufacturing Company's works, has a plan in his mind, and is now at work on it, having this end in view. Breaker cards, as a general thing, need more stripping than finishers—both tops and cylinders.

With regard to stripping cylinders, it continues to **be** done mostly by hand; but **we** strip very much less than we used to do in former years. The first the writer heard about any change was some eight or nine years ago, when an old associate called on him who was then running a large room in an adjoining State. We had not met for years, and this was a very agreeable meeting; and, in a hurried manner, each asked and answered a great many questions. Among others **he** asked, "How often do you strip cylinders?" He was told once in two hours; that was the old style, and was considered the orthodox plan. "Well," said he, "try stripping them once in four weeks, or as often as you grind." He was supposed to be jesting; but he added, "I am in earnest; and that is as often as I strip mine." It is needless to stop to describe how much astonished the writer was to hear such talk, from such a source; but he had long ere that day made up his mind to try whatever he heard of in way of improvement if it seemed reasonable, and sometimes if it did not. This was one of the times when it did not so seem; but directions were given a stripper not to strip a certain cylinder any more until further orders; and it ran without stripping until the time of grinding, some three weeks, and no perceptible difference was discovered in the work of that card. It was watched daily—almost hourly. One day, while the carder was at his desk, marking down time, near night, a stripper came up and said he was desired to say to the overseer a certain grinder wanted to leave. . He was asked what the trouble was. "O," said he, "he is on a job he thinks he will never live to finish, and he wants to leave before he is 'killed entirely.'" "Well, what is he doing?" "Stripping that cylinder that has run so long without stripping." The cylinder was examined, and there was plenty of time to examine it before the bad job was completed. It stripped hard—very hard, so hard indeed that the carder was fearful that the clothing would be injured before the stripping was completed; and it was thought it had run too long without stripping. No others were tried for the same length of time; but they were tried at shorter periods until *one day* was settled down on as the proper time for a cylinder to run without stripping. The writer has adopted that plan, and practised it for a number of years, having one-half the cylinders stripped in the forenoon, and the other half in the afternoon—every other cylinder in each row. Perhaps cylinders might run longer than that, under some circumstances—as for instance where the work is light and the cotton clean; under other circumstances it might not be often enough, where the work is heavy and the cotton dirty. No positive rule can be laid down that will be safe to follow in all cases. The carder must exercise his own judgment. But there are two or three things which are pretty certain: first, the cases are very rare where it is necessary to strip cylinders once in two hours, as we used to do; second, the cases are rare where they can be profitably run four weeks, or three weeks, without stripping; third and last, but not least, all kinds of stripping, whether cylinders or tops, and whether they are stripped often or otherwise, one-half should be stripped at a time—every other top on a card, and every other cylinder in the same row, for the reason that every stripping makes the work lighter for the time being, and where alternate tops or cylinders are stripped, it affects the work less than if they were all stripped at about the same time.

The question of placing screens under cylinders, or not placing them there, or anywhere else, has been considerably agitated at different times and in different places. Some use them; others do not. They had a very general popularity once. Now they are popular in some places, and in others they are not. The author has used them, has nothing to say against the principle, rather likes it, and if screens are properly made, of the right material, he has no doubt they might be, and perhaps are, a benefit. Screens made of wire netting did not work satisfactorily with him; there was too much of a tendency to bed over and stop everything going through them; then they were no longer screens, but a sort of under-casing, that keeps too much dirt in the work. Cleaning often enough will obviate this difficulty; but it is necessary to do it so often it becomes burdensome. Zinc, with round or other holes punched in it, is objectionable, because it is so difficult to make the edges of the holes smooth that it is hardly ever done perfectly, and the result is cotton dirt clings to them and soon stops them up, though they are not as bad as wire netting. A cast-iron rack, made of bars and set in grooves that conform in shape to the surface of the cylinder—the bars made three-cornered, flat side next to cylinder, and two sharp edges forming the edges of the opening, set three-eighths of an inch apart, and the other sharp edge down, and the general surface of the rack five-eighths of an inch from the cylinder—works well under some cylinders. There is probably no reason why such a rack would not work well under a main cylinder of a card if placed there. Such a rack would not probably get clogged up very easily.

Any kind of screens now in use can be run advantageously, if sufficient care is taken to keep them clear. One very common error is placing them too near the cylinder surface. They should not be placed less than five-eighths of an inch from it. If waste from the cards is to be re-worked at all there should be as much of it kept in the work, as it is passing, as is possible to keep in; or, rather, it is best to keep the fibre in, and let the dirt out. Screens will do that on common kinds of cards.

The speed of main cylinders varies considerably under different circumstances and as managed by different men. If there are no screens under main cylinders one hundred and twenty-five revolutions per minute for a 36-inch cylinder, or in that proportion for other sizes, is about right. If such cylinders run much faster than that they throw off too much waste; if they run much slower they do not clean the cotton well. If there are screens, they may run as high as a hundred and fifty or sixty, without harm, and they clean the cotton better on high than they do on low speed. The surface of the main cylinder should exceed the licker-in speed from twenty-five to thirty per cent.

The draught of cards varies as much as speed. They will draw as much as they are asked to do, as a general thing; and yet it is not best to go to extremes in either direction—not too low nor too high; from sixty to eighty, as a general thing, is about right for the draught of a card of ordinary dimensions. If we have a thin lap and consequently must have a low draught, it necessitates a slow speed of the doffer, and the cylinder will not clear well. If, on the other hand, we have a very thick, heavy lap, making it necessary to have a high draught, the lap being so thick is not held firmly enough between the rollers to be well carded. It will draw off in flakes, more or less; so it is best to have a medium draught, if possible The idea that cotton passes round a cylinder several times, before being taken off by the doffer, is fully believed in by many carders and spinners. But the author of this work is of the opinion that the cotton that passes the doffer once is either thrown off under the cylinder in the shape of flyings, or it sticks to the main cylinder and remains until removed by the card of the stripper in the shape of main cylinder waste. If the feed and doffer be stopped for any length of time and the cylinder and the top stripping goes on, the doffer will receive a little additional streak of waste, and the cylinder will throw some of its waste upon the newly stripped tops; but when everything is in motion, what the doffer does not take off when it first reaches it, does not come at all in that way. If any one wishes to see how cotton looks, and what kind of a condition it is in, after passing by the doffer several times, their curiosity can be gratified by drawing the doffer off from the cylinder a thirty second of an inch, and allowing it to run in that manner a few minutes. But it will be at the expense of the cotton so carded. It can be seen in another way: Grind the card hard enough to barb the points of the teeth so that the cylinder will not clear, and the same result will follow. The fibre will be ruined in either case. The beauty of a card of any kind is to have it so arranged as to readily seize the cotton and as readily let it go, when the proper time comes; in other words, it should clear well—nothing be allowed to load except tops, and allow them to load with dirt, leaf and other foreign substances instead of cotton.

"How often is it necessary to grind cards?" is a question frequently asked. Well, the question can be asked quicker than it can be properly answered. As a general thing they should be ground often enough to keep them in good working order; but that is very indefinite; and another question is immediately asked—"Well, how often is that?" That depends entirely upon circumstances. If a card is properly ground and adjusted, no one part rubbing or chafing against another, and it is carefully managed, not over-loaded, and good stock well prepared in the picker, it will run a month without grinding, or between grindings, very well. But if ground once in two weeks they would work better; or one week, better still. Indeed it would be difficult to grind too often, if the grinding is properly done. There is no time a card works so well as it does the first day after it is ground.

When we used to have fewer cards, card heavier than now, on the same kind of work, we ground oftener, and it was necessary to do so, too. A grinder in those days used to have 26 cards to grind; and he got around with them once in two weeks or thereabouts, generally grinding two cards per day. Those were the old wooden-framed cards, which were not as reliable as iron frames or as easily managed. Nowadays we have modern cards, card lighter than formerly, consequently have more of them; and a grinder has from 50 to 60 cards to take care of, and he should grind 15 per week; and that usually keeps them in very good condition.

In taking care of cards much depends on the thoroughness and skill of the grinder. If when he sets a card up, he does it in such a manner that everything remains in place until the time comes to grind it again, it will be in a comparatively good condition. If, on the other hand, things have been left loose—not properly adjusted, and those that have been properly adjusted not properly secured, and move—the doffer, rolls, or tops get on the cylinder, the points are soon spoiled, and the card will not work satisfactorily. To be sure, the cotton goes through; but it will not be carded: it will be *ground* through. It is impossible to make good yarn of cotton which has not been properly carded. No future process, either in the carding room or the spinning room, will remedy such an evil as that.

How often to re-clothe grinders is another question. Some carders allow their fancy or Hardy grinders to run a long time; some a year, and others not so long. The writer is of the opinion that a fancy five-inch grinder should generally be re-clothed once in three months, and a twelve-inch top grinder once a year; a Hardy traverse at least once a month. A grinder should not be allowed to run so long as to get glazed over. If they do, they are of very little service, and will not grind satisfactorily. The regular motions may be gone through with, but a card will not grind well with an old, smooth, glazed grinder; and if it is not ground it will not work well.

Whether to re-use old emery is another question. It is contended by some that emery will not come the same side up two different times—that it is just as well to use it a second time. But let us see: Is it not just as likely to come the same side up twice, as it is to change sides? Of course it is; and for that reason, if for no other, it should not be used a second time—so it seems to the writer.

How long shall we run old card clothing, is another important question. A cylinder has become short, some teeth are broken out, tops have got worn short, or a doffer is broken out in spots. At what stage shall it be renewed? As soon as it begins to work badly. It is not profitable to run card clothing so long that it makes bad work—not even one card, for one bad-working card will often injure the work of a whole section. But a good manager may use his old clothing all up by taking a little pains. When a majority of the sheets in a main cylinder become bad, remove the whole and re-clothe with new sheets; but pick out the tolerably good sheets and lay them by. When another main cylinder is found with a part of the sheets in a bad condition, remove the bad ones and replace with those that have been saved out of the last lot. It is necessary, however, to exercise great care in the selection of sheets to patch up with. Those longer than the original should never be used; but if they are shorter, one or more strips of cloth or paper may be laid under them, stuck with thin glue to hold them in place until the clothing is put on. A gauge or card calipers may be used to ascertain the length of teeth accurately. Tops may also be patched in the same manner, without the necessity of measuring the length of the teeth, as they can be set to accommodate different lengths without difficulty, so far as carding and grinding is concerned. It is necessary, however, that tops should all take into the strip card alike. To accomplish this a nice little arrangement has been made by placing two set screws in the plate; the heads are slotted and may be adjusted by a gauge made for the purpose, so as to work on the jaw of the stripper and allow the top to take into the strip card exactly right—no matter whether the teeth are long or short. This is the invention of Jabez Edwards, a veteran carder on the Merrimack Corporation.

The speed of the different parts of a card, so far as rollers and doffer are concerned, depends upon the draught. The speed of a main cylinder may be the equivalent of 125 revolutions per minute for a 36-inch main cylinder when tops are used, and 150 with workers and cleaners, or workers and strippers. It is a fact that main cylinders are run much faster than the above in many places, and it is claimed that they work well, even better than when run at a lower speed; and it will not be disputed. The writer has tried high speed for main cylinders, and has no fault to find with their work, but they throw off a great quantity of waste where there are no screens, and they get a terrible battering when anything gets into them beside cotton, as will sometimes "happen in the best-regulated families."

There should be just draught enough between the lap roll and feed roll to keep the lap straight. Say two-hundredths of one or one and two-hundredths, as figures would express it. The same between the comb and calender rolls, where there are such, or between the comb and apron, where there are no calender rolls.

The principal drawing should be done between the feed rolls and doffer; it can be done there to almost any extent without injury; but when cotton is passing in a sheet or sliver any considerable distance *it should not be drawn.* But this subject will be considered more at length under the head of drawing and speeders.

Cards, like everything else that is used much, need washing sometimes. All the iron work and the lower wood work may be washed with soap and water; but water should never be applied to tops, as it is very apt to cause them to spring. Rub tops with waste made damp with benzine or spirits of turpentine; afterwards rub with waste made damp with raw linseed oil. When done with both kinds of such waste either burn it in the fires under the boilers or throw it in the river, as it is very unsafe to be kept with other waste. Never try it; for it may take fire.

A well-arranged carding engine is almost as much of a marvel in its operation as a modern picker. It takes the cotton as left by that machine in the shape of a lap or sheet, receives it very slowly and turns it out in the shape of a sliver or continuous roll, and very clean. This is, in fact, the last regular cleansing process. It is true that some dirt rattles out of it as it passes from one machine to another, and through them. But the great cotton-cleanser is the card; all the little nits, or neps, that cling to the fibre through the card continue to cling to it until they reach the cloth; hence the great importance of keeping cards sharp and in good working order all the time, for no other machine that succeeds the carding will do the work of the card.

There are various kinds of cards in use in different places, and something is claimed by different men who have from time to time improved upon Arkwright's and Slater's cards; but that those gentlemen made so long a step in this particular at the beginning, that less has been done by way of improving cards, within the last twenty or thirty years, than has been accomplished in other departments, is a fact well known to all acquainted with cotton manufacturing. The self-stripper is about all; otherwise we are doing the carding by about the same methods as it was done thirty years ago. We don't card quite as heavily, but the general plan is nearly the same. If some cotton-working Rip Van Winkle, who was laboring in a mill forty years ago, had fallen into a sleep and should wake up now (October, 1875) and be placed in a dressing-room, he would hardly know where he was, or remember that he had ever seen anything like it. Put him into a spinning-room, and he would probably remember a little more. He would of course know the weaving by the noise. But place him in the carding-room, and he would feel at home. Railways and the present lap heads would look a little odd, but the familiar look of the cards would set him right at once.

There are two principal cards now in use: the card with tops or flats, as they are often called, and the roller cards, or cards without flats, and instead small cylinders, called workers and strippers or workers and clearers. Both kinds have very strong advocates, but the former are the most numerous. Workers and strippers have been placed under the main cylinder by some, as well as on the top. But it has been found by actual trial that they don't work very well. Within the last year two gentlemen of this city (Lowell, Mass.) have been studying on an improvement in carding; in other words, on an improved card. They have succeeded in bringing out a machine that certainly promises well. It is a common 36-inch main cylinder, but instead of being placed in a square frame, like a common card, it is placed on a kind of pedestal or foot, so that a large portion of the cylinder underneath can be seen and easily handled by a little stooping. The doffer is placed a little higher than in an ordinary card; the feed is under the doffer. There is a regular licker-in, and between it and the main cylinder is another cylinder, a little larger than the licker-in, which takes the cotton from the licker-in, and the main cylinder takes it from that. Under both these little cylinders there are racks, in the form of screens. The licker-in rakes the cotton down, as it takes it from the feed rolls. The main cylinder rakes down as it takes the cotton from the second little cylinder. Underneath the main cylinder, from the point where it takes the cotton from the second little cylinder, is placed what would generally be called tops, but in this case it would be more proper to call them flats, twenty in number, and an equal number on top, reaching over to the doffer, from which it will be seen that this card has double the usual amount of flats and an extra cylinder. This seems to be an improvement. It is certainly most skilfully arranged, taking advantage of the laws of nature in getting rid of dirt—throwing it *down* instead of carrying it up and over, as is the case with an ordinary card. There is a small space between where the second cylinder delivers the cotton to the main cylinder and the under set of flats. A rack or screen has been placed there, under the main cylinder; but as this cylinder acts altogether differently from a common one, inasmuch as it carries its whole load down, the centrifugal force drives a portion of good cotton down through the rack, and it must be a very fine rack to work there. In a common card the main cylinder takes the load up under the tops or rollers, as the case may be, and then delivers it to the doffer. A screen or rack may be placed under such a cylinder as that, and as the cylinder is comparatively empty, not much good cotton can be thrown through it. But it will readily be seen that this is a different affair. Twelve of these cards were put into No. 1 Carding-Room, on the Merrimack Corporation, of Lowell, Mass., a few weeks since, and put to work by the side of twelve finishers. These cards are supplied with laps from the picker, and are consequently what may be called single cards, or single carding. The work is being kept by itself, for the purpose of testing and comparison. They have not been running long enough yet to enable one to come to any definite conclusion as to their real merits. But they certainly promise well thus far. It puzzles the oldest heads to pick out the drawing or roving. The spinner reports that the work runs well, and the yarn tests well—a little stronger than the double carding. What it will do in the way of preparing material for good, smooth, clean cloth remains to be seen, although there is no doubt but they will do much better business than ordinary single cards.

The writer has been somewhat prejudiced against single carding for some years, having had some unpleasant experience with it during three years, at one period of his life. Trying to card too much to a card was what did the mischief in that case; and the same complaint has troubled many another man, and other concerns besides the one where this occurred. The trial has often been made on this wise: to increase the work of the same kind of a card one hundred per cent.; that is, make one card do the work of two. It has almost always been found to be too much to work well. But the kind of card just spoken of is new and has many advantages over the old ones. First, the arrangement for feed is very much better—the licker-in and main cylinder both working down; second, nearly the whole surface of the main cylinder is at work all the time; third, there are two entire sets of flats, half of them under the cylinder instead of on top in which position they are much more effective. There is an arrangement by which the bottom flats are stripped twice to the top set once, though the stripping is all done by one set of double-acting cams. It is claimed by the builders (Foss & Pevey) that this card will do seventy-five per cent. more work than an ordinary card, and equally as well; that is, in the proportion of four cards to seven, whether carding single or double. There is at present no one prepared to dispute the claim. If this claim is met, or the expectation of the builders realized, it will be seen that a great saving of floor-room will be made as well as power, appurtenances and help to run the machines, though of course help would not be reduced in regular proportion. If they do more work, it is more work to tend them, and they must also be ground oftener to keep them in good order. But this card is a long way ahead of the old kind, as it is now; and it is not completed yet. The builders, and those who may have charge of them, will think of a great many improvements that will be suggested from time to time by working with them.

This card has a decided improvement in the way of driving the stripper. It is so arranged that the stripping starts and stops with the feed; when the feed stops the strippers stop, too, instead of working away in the old manner until they have nearly emptied both tops and cylinders, causing a breakdown and light work when the feed starts again, as they always do on the old plan, if the cylinders run any length of time with the feed and doffer standing.

It is more than probable that the main cylinder might be run as high as 150 revolutions per minute, instead of 125, as at present, especially if there is a solid casing put under it, between the second little cylinder, instead of a rack, as it seems likely there will be. It is possible that this is the grand central point where the advocates of both double and single carding may harmoniously meet, this being not exactly the one or the other, but a sort of compromise, or cross between the two, like Cropton's mule.

Whatever kinds of cards may be used, it is necessary, in order to secure good work economically, to have good, nice card clothing. Perhaps the writer may be allowed to digress from the main object of this work enough to say that the Lowell Card Company make such, and he speaks advisedly, as he has used this Company's clothing for some years, with entire satisfaction.

Another important qualification for good carding is good grinders: machines are meant. The Ladd grinder, built by Mr. A. H. Saunders, of Nashua, N. H., is a great improvement on the long fancy grinder, as the traverse is positive and can be regulated to the wish of the operator, is reliable, more comfortable to handle and does better work than the old-fashioned, haphazard traverse. The Hardy grinder is good and makes a very nice-pointed tooth. Any grinder must be kept in order and should be perfectly true and straight.

Since the author commenced this chapter he has made further experiments in grinding. In a couple of stands, rigged up at the ends of the doffer and nearly over the comb shaft, is a little shaft one inch in diameter, driven from the main shaft overhead. From this shaft the main cylinder is driven 44 revolutions per minute, the grinder 400 and a doffer 124, or thereabouts. In this manner a card is ground very nicely; with any kind of a good grinder they grind quicker and better, and it don't seem to be so hard on the clothing as the old style.

Where breaker cards are used, and consequently lap heads, it should be the aim of the carder to so guide the different grists from railway boxes into the frame of the lap head as to make a uniform, even lap for the finisher cards. If one laps over the other and there are thick and thin stripes in the lap, the finisher will not work well. The cylinder will have stripes in it; and the teeth, where these stripes form, will break out very soon.

Since the preparation of this chapter was begun the author has invented another new card guide. It is cheap, convenient, durable and can be placed in any position within range of the railway box in a quarter of a minute. There is no patent on it. Anybody can make and use it who chooses to do so; and another splendid thing about it is, it does not infringe on any patent. Take a piece of nail plate, three-quarters of an inch wide and long enough to reach from the calender roll to the top of the railway box; turn a knee or elbow, an inch and a half long and running horizontally close to the top of the railway box; to this short piece rivet a common card trumpet or eye, such as is used in the cover of a railway box; drill a hole in the upright part, within an inch of the top; suspend the whole thing from the calender roll-stand with a five sixteenth bolt; let the bolt hold it in place. In this manner it will swing the entire width of the railway box, and can be secured at any point by simply tightening up the little bolt. This is a good guide. The only objection to it is, it leaves a hole open across the box; but the objection is in proportion to the size of the hole, so make it small; then the objection is small. It is of great importance to have the ends from the cards uniformly distributed on the railway aprons of both breakers and finishers. A cheap and convenient guide is necessary to do this. It pays to manipulate them as often as cards are ground or anything happens to disarrange the ends on the apron.

There are a great many different opinions as to the best method of disposing of waste made in a carding-room. Some advocate throwing it out of the work altogether, where the work is fine, or where it is desired to have it nice, and work the waste by itself into some coarser material. This plan might work well; no doubt it is practised in some places.

The writer has some few suggestions to make at this point: First, let the construction, arrangement and management of carding engines be such as to make no waste fit to work over for any kind of cloth on these machines. Let cards have screens of some kind, either such as are suggested in a former part of this chapter, or some better ones, allowing a space for dirt to drop out, but no fibre, or so little fibre and so short that it will not be worth working over for anything. That plan takes care of flyings. Second, cut out under the feed rollers of breakers so as to let dirt drop out there, before it goes any farther; quite a quantity will drop out there. Apply the mote-catcher to finishers, and empty them as often as they get full. Third, so far as dirty waste is concerned, don't strip either tops or cylinders, on any kind of cards, often enough to remove any waste that will be fit to work over in any kind of cloth. Let cards run without stripping until the waste is dirty enough to go for batting. Cards will fill up to a certain degree with good cotton immediately after they are started; then they may run a long time—*much* longer than is generally supposed—without taking on more good cotton, but will continue to catch dirt.

This plan provides for dirty card waste, such as strippings and flyings, and such as is generally worked over with the cotton. Such waste can never be mixed with cotton in so nice shape as it is in while already there. Why not keep it there, instead of knocking it out, carting back and forth, re-mixing and re-working, to the great injury of the whole work? For the management of clean waste, the reader is referred to another part of this chapter. The writer has only one amendment to offer to the plan suggested there; that is, instead of running it through a finisher-picker, as is there recommended, let a new picker be built on purpose for it, with one beater and one set of cylinders. Get it back into a lap as quickly and easily as possible.

One thing more in this connection: Let every carder see that no more waste is made at any point in his room than is absolutely unavoidable. Every pound of waste made is a dead loss, so far as labor is concerned; and moreover it is a positive injury to the whole work when it comes back. Arrange machines so that they will run well and have them watched. Nice carding is the soul of good yarn.

CHAPTER IV.

Drawing.

GENERAL REMARKS — RAILWAYS — ENDS IN THE BOX — TWO SYSTEMS — GUIDES — WHERE TO DRAW AND WHERE NOT TO DRAW — THREE GENERAL RULES — HOW MUCH TO DRAW BETWEEN BACK ROLLERS — EVENERS — THE WHOLE DRAUGHT OF A RAILWAY — AN EXCEPTION TO THE RULE — WEIGHTING TOP ROLLERS — DOUBLINGS — SPEED OF FRONT ROLLER OF RAILWAY — ELECTRICITY: HOW TO GET RID OF IT — TOP ROLLERS — SHELL ROLLERS — SPEED OF FRONT ROLLER OF DRAWING FRAME — CROWDERS — DOUBLING AND DRAWING — PHILOSOPHY OF DRAWING — TRAVERSE — CLEARERS — WEIGHING DRAWING — BELTS — GENERAL SUMMING UP.

DRAWING succeeds carding, and very properly, too. The object of drawing is two-fold: To straighten and lay the fibres of cotton parallel. Whatever a picker or cards do to cotton, there is one thing they do not do, and that is, they do not straighten the fibre much. If a fleece of cotton is taken up and held between the eye and the light, it will be noticed that the fibres lie almost every way. There is, perhaps, a slight tendency of a majority to lie a little more in one direction than another; but as for anything like a general straightening, it is not found there. Again: as cotton comes from cards, the different slivers vary in weight considerably, owing to a number of causes: First, the laps from the picker may vary a little; second, all cards do not clear equally well; third, there may be a little variation in the feed rolls, doffers, doffer pulleys or calender rolls. (See Chapter I.) From all these causes there is a variation in weight, and hence the necessity of doubling; and this brings us to the second part of drawing, viz: To draw it down after being doubled. When the doublings of a mill are counted, we begin back at the finisher-picker, and it is proper to do so; but so far as the evenness of the work is concerned, doubling before it comes to the drawing frame, as the saying is, "don't amount to a row of pins." It is a most excellent way to mix cotton, and good mixing is indispensable to uniform yarn; but so far as regulating inequalities of weight in the different slivers, it does not do that to any great extent. Drawing is a very important branch in the preparation of cotton for spinning; and in order to secure good, even yarn, all the drawing frames must be properly built, well-proportioned, rightly adjusted, kept in order and carefully tended.

There have been, and are still, a number of different kinds of frames in use, each possessing some merit. These frames have been greatly improved within the last twenty years or more. When they were first brought into use there were no stop-motions attached, and if the girl who attended them did not see an end when run through and so stop the frame, it continued to run just as well with a portion of the ends out as with them all in, but the work suffered. If she did not take out the single, of course there was a portion too light, and if she did, then waste was made. After a time stop-motions were applied to the frames in such a manner that when an end broke or ran out the frame stopped, and so they continued to run for years; but within a few years past the stop-motion has been extended, and now if one of the strands breaks in front, or if a bunch comes down in front, the frame stops. The same principle has been applied to railways, and it is a great advantage, as that machine cannot stop long at a time without great inconvenience and loss.

Drawing frames have been built in various shapes and sizes. We have them all the way from two to six lengths of rollers, and from three to six sets. We find them with plungers, cast-iron crowders, coilers, wooden and tin crowders, and no crowders at all. They are running with dead weights and lever weights. Cans are made to revolve in some machines, and in others they are stationary, and they vary in diameter from six to twenty inches. There is a great variety of opinion as to the proper speed to run them; how much weight should be applied to rollers and how to apply it; how much to draw and where; how much to double, &c.

We have already considered the nature of cotton; it is not necessary to repeat here. We have traced it in the various processes through which it passes in preparation down to the railway box or the can, as the card has delivered it. We find it as it leaves the card in a light, loose and almost transparent sliver—tolerably clean, but fibres lying in every direction. The opening and cleansing has been performed. The next thing in order is to lay the fibres parallel, in smooth, compact and even sliver. For this purpose it is drawn out, doubled and drawn several times, according to the quality of the work desired, the taste of the manager, or both. Railways have been very generally adopted in this country within the last twenty-five or thirty years, and this is the most convenient, economical and comfortable method of gathering a number of card slivers together in shape to receive the first drawing; and this, too, leads us to a consideration of the railway-head. And, first of all, it is necessary to deliver the slivers to this same head in a smooth, even sheet. The edges should not be allowed to rub against the box in their passage through it; if they do, they will get fretted and will not make a smooth, even-edged sliver of drawing.

There are two systems of laying card slivers into a railway box. One we will denominate the B system; the other the E system. The former is to make a calculation of how many thicknesses of sliver from cards it is necessary to have for a certain width of apron, in order to cover it uniformly, and whatever number it takes lay that number down, one sliver exactly over another, in the center of the railway box at the farthest end from the railway; the next two ends meet in the center of those already laid down; and then shingle on each side alternate ends until they are all used. This system makes a most excellent sheet, and its only fault is, it leaves the edges the thickness of a single card sliver. But it has advantages over all others: first, it makes the best sheet, with the single exception mentioned above; second, it retains its shape the best, when one end is out by reason of a card stopping to be stripped or for other purposes. The tendency is to close up a little, and so partially cover up the deficiency, making the sheet narrower instead of leaving a vacancy.

The E system, like the other, needs a calculation as to the number of ends and width of apron; when ascertained, lay down one-fourth, one-third or one-half, the ends from the first cards or those farthest from the railway, till the apron is covered; then repeat layer upon layer until they are all used. This system makes a very nice sheet and square edges; but they are apt to spread out somewhat and chafe the edges against the sides of the box—especially those first laid down. Again: when one card is out, there is a square deficiency; no help for it. With a perfect, adjustable guide either system can be used, according to the taste of the operator. Without such a guide, it is difficult to arrange satisfactorily or employ either system.

There is one other system at work in some places: the haphazard system, where the ends from cards are laid on the apron without regard to position—in some places four double, in others two, and others one, just as it happens. Such cannot make good work long at a time, as they soon spoil the rolls in the railway-head.

The card slivers, after being nicely laid down in the box on the apron, are ready to pass through the railway-head; and at this point, let it be remembered, cotton receives its first doubling and drawing proper; in other words, it is the point where the doubling and drawing begin to tell on the evenness of roving and yarn. It is well, in making calculations of doublings, to begin at the finisher-picker or lapper; and it is as well, also, to bear in mind that, so far as the evenness of roving is concerned, these doublings don't amount to much until we arrive at the railway-head. The others are good in their place. We cannot dispense with them. But it is at the railway-head where we begin to draw down and even our work. A railway generally has four rolls, or rather sets of rolls; at least all should have this arrangement. There has been a great deal said about the right number of rolls, or what number is right, for a drawing frame; but not much about railways. The principle is the same in both frames. The difference is, the sliver or sheet is heavier in a railway than in a drawing frame and consequently takes more power to hold and draw here; and somebody probably made the discovery, at some time, that four sets of rolls would do this work better than three. He did well: they do ; and it is probable that if all who have used railways from that time to the present had known as much as the man who made the discovery, there would not have been so many mistakes made in arranging the draught of the machine.

As it is, we find them drawing considerably between each set of rolls. We may safely conclude that the man who found it necessary to have four sets of rolls in a railway did not draw very much between the second and third sets. He did not place them there for that purpose, but to hold the sheet firm while the first and second pairs of rolls did the drawing. This may seem to be a strange doctrine ; and perhaps just here a few remarks on drawing in general may be in place. It is a delicate and somewhat complicated subject to handle, and the writer may not be able to make himself perfectly understood, but he will do his best.

First, cotton fibres, as we find them in slivers of drawing, are somewhat curled up—more so in card slivers than any other, from the fact that every process of drawing, proper, straightens them a little ; second, the object of drawing is not only to straighten the fibres and lay them parallel, but to draw down as well, especially after doubling. What we want is the best method of doing this business of drawing fibres straight, laying them parallel and reducing slivers down to their proper tenuity.

There are three general rules that may be observed with profit, though like all rules there may be exceptions to them : first, do the principal part of drawing in all kinds of drawing, roving and spinning frames between the first pair of rollers, or the two first sets ; second, draw no more in any one place than is strictly necessary ; third, increase as it progresses or advances from one frame to another on its way toward yarn. And now we will explain the above rules, look at them in their true light, and mention exceptions.

Perhaps the writer can make himself understood as well, if not better than in any other way, by introducing a little incident. Some years ago a gentleman called on him and entered into conversation concerning the preparation of cotton for spinning. He said he had been brought up a mule-spinner, but was then a mill owner and cotton manufacturer. He professed to understand the whole process of preparation, in all its particulars from beginning to end. We talked about drawing, among other things: the best way to do it, where to draw, how much, &c.; and in the course of his remarks he said he would have only two sets of rollers in a drawing frame, if he was going to build them for his own use; that the third set were useless; that the drawing should all be done between the two first sets, any way, and what is the use of a third set? The writer took the matter into serious consideration. He asked himself the question, "Of what use is the third or fourth set of rollers in a drawing frame?" He agreed with the ex-mule-spinner, that the most of the drawing should be done between the first two sets of rollers. But why are there three, four and sometimes five sets? Have manufacturers and machine builders been throwing all this labor, material and power away? It could not be possible; and so he set himself to work to find a use for the third and fourth sets of rollers in a drawing frame. He thought like this: Suppose we remove the third set of rollers and do all the drawing between two sets; then we must weight the back roller heavy enough to hold the sliver while the front rollers draw it out. In a word, we must put as much weight on the back set as we now have on two sets (if they are rightly adjusted), and that amount of weight would soon spoil the top rollers of the back set. But if we can divide the weight necessary to hold the sliver while it is being drawn among two or three sets of rollers and make them all hold, then the weight of each individual set of back rollers will be comparatively light, and the top rollers will keep in good condition and last a long time. Once more, by way of illustrating and enforcing this first rule: Pick up a sliver of drawing, take it between the thumb and forefinger of both hands and grip it in two places; let there be a space of a half inch between the two grips; pull apart a little, gently, and it will be perceived that there is a little give or yield before separation takes place. That little give is taking out the curl of the fibre, straightening it preparatory to drawing out, or a separation of fibres and attenuation of sliver; and that give or yield is the exact amount of draught required between the two or more sets of back rollers; no more is needed, and less would be about the same as none at all. In this manner all the sets of back rollers hold, whether two or more, and the front set does the drawing, and that is right.

But somebody says—"I draw nearly as much between my middle and back rollers as I do between the front and middle sets. I divide my draught pretty nearly equal on drawing frames." You are not the first man who has done that same thing, and probably not the only man who is doing it now. But there is a more excellent way to draw cotton sliver. The construction of drawing frames seems to say very plainly that they were not designed for such use. If it is proper to draw as much between middle and back as between front and middle, then most certainly they should be set as near together, which is not the case. Again: if we undertake to equalize this draught, we must weight them all equally; and more than that, we must put more weight on each set than we are called to do when it is drawn according to rule, for it will draw harder in both places. Another man says—" I do not draw the same between both sets of rollers, but I draw some between the middle and back sets—considerably less than between the front and middle. I think that the best way." So it is, if the "some" is small enough; but if you draw enough to separate fibres there, it is very hard work —hard for rollers—and it will not draw even; it cannot; the rollers are too far apart; it won't come out regular, but in flakes. Draw just enough to take the curl out of the fibre, and you are right.

The question now arises, How much should we draw between back sets of rollers? What should the draught be, as shown by figures? Well, that depends somewhat upon circumstances; it depends upon how much the fibre is curled up. We find it most curled in card slivers, and therefore should draw more between back sets of rollers in railways. And this takes us back to where we started. The other two rules will be explained as we progress in this chapter. We may allow a draught of from 1.25 to 1.50, as figures show (see remarks on draught, Chapter VI.), between the two back sets, and from 1.15 to 1.25 between the middle sets on a railway-head, and do the rest of the drawing between the two front sets. We may draw more in these places on a railway than a drawing frame, for the reason that the fibre is most curled there. The whole draught on a railway, where eveners are attached, is necessarily variable, and most railways have them in these times. They are a very important and useful appendage. Some object to them on account of the constant changes in draught; but let it be remembered that we are obliged to have these changes in draught or changes in weight. The former are preferable. As railways take the slivers of a section of cards varying in number from six to twelve, and as there is more or less stopping of cards for stripping, grinding, accidents and laps running through, without eveners it is impossible to avoid light work unless it is broken and taken out, and that operation makes a great deal of waste, which is always to be deprecated. There is an arrangement through a double set of draught gears by which the absence of the end from one card is made up; but it is only one. If two ends are out the work must be light, and nothing seems to remedy such evils except the evener. That is not all the benefit of the evener: it acts in the other direction, also, giving notice of, and counteracting heavy work, though there is not usually so much trouble with the latter as the former.

The proper point at which to regulate an evener belt on the cones is about one-fourth the distance from the large end of the driving cone; and the cones, and the whole evener work, should be of sufficient compass to counteract the absence of three ends from the cards. The whole draught of a railway may vary from three to five where there are eveners, and where there are none it may be set at four and a half. But here is an exception to the third rule laid down, viz: to increase as it progresses from one frame to another. A railway should, as a general thing, draw more than the first head of drawing proper, for the reason that the sliver is heaviest and the most curled state of the fibre exists there. The weight necessary to apply to the top rollers of a railway, or drawing frame of any kind, should be exactly enough to hold, not a pound more. In many modern railways there is a rack by which the weight of all the rollers is connected and the power applied with one long lever and one weight. This is a good arrangement. Let the carder experiment with the weight until he "touches bottom," or has reduced weight until it will not hold to draw; then increase until it will hold.

We should have no more power applied to the top rollers of a railway-head or drawing frame than is absolutely necessary, for three reasons: First, an extra amount excites more electricity than the right amount, and there is enough of that under the most favorable circumstances; second, it is an unnecessary wear and tear of top rollers; third, it requires more power to drive it. There should be no draught between the front set of rolls and the calender rolls, either in a railway or a drawing frame.

It used to be the rule to have a little draught in that place; it was supposed to be necessary. In a drawing frame the front fluted roller is usually one and one-quarter inches in diameter, and the calender roll two and one-half; the old arrangement was the following: Suppose there were 36 teeth in the driving gear on the front roller, then it was thought there must be a gear of 71 teeth on the calender roll, giving the thin, gauze-like sliver a slight draught. A little reflection will enable any one to realize the impropriety of drawing cotton in this place. First, the long distance between these two sets of rollers—usually from ten to thirteen inches; second, the condition of the sliver, so thin and delicate that a breath will sever it; and when we come to apply a continual strain to it, in the shape of a draught, it must necessarily make it uneven. There is usually contraction enough in the fibre of cotton to keep it straight in this place while running exactly square—that is, no draught at all; and when there is considerable electricity excited and the air is dry, a draught here becomes intolerable, is exceedingly difficult to manage and materially injures the work. The writer is aware that railway-heads and drawing frames are built and arranged for a slight draught here, and that the general opinion of machine builders and carders is that it must be so, that the frames will not run without it; and this is an old, deep-rooted and long-settled conviction; probably nothing but an actual trial will dissipate this illusion. If the rolls are accurate, the front exactly one and one-quarter inches, and the calender exactly two and one-half inches in diameter, then there may be a 36-teeth driver on the front roll and a 72-teeth driven on the calender, or in that proportion for any other number of teeth, and it will run nicely. The author has run drawing in this manner for years; he knows what he is talking about, and he claims to be the discoverer of the draught error, and the author of the "new departure." A trial will satisfy the most sceptical of the correctness of it.

The doublings proper commence with the railway-head; the number of course depends on the number of cards that run into one head; it may be six, nine, ten, twelve or fourteen. The speed of the front roll of a railway is necessarily variable where there are eveners attached, as here is the variation of draught. Medium speed of a front roll one and one-half inches in diameter may be about 370 revolutions per minute; maximum,

390. The above kinds of speed will answer very well where there are stop motions attached, but where there are none, they should run slower, as when railways run very fast, without stop motions, when the front roller rolls up it is liable to do a large amount of mischief in a short time.

Speculations have been indulged in as to why the front roller of a railway and drawing frame is larger than the other rolls. The inquiry has been raised, "What is the object? Are they any better than they would be if the rollers were all of one size?" The author never heard any one say aught in answer to these questions; but he will venture a guess, based on his knowledge of such things in general. The reason for it is probably to save running the front roller as many revolutions in a given time as it would be necessary to do if they were all alike. There is no particular objection to running a railway or drawing frame at a high speed, if they are provided with modern stop-motions. The work is as well done on high as low speed, notwithstanding there are many of a contrary opinion. The writer once took charge of a room where there was a superabundance of drawing, and he thought it would be a good time to improve his work by diminishing speed of drawing; for that purpose he reduced the speed of one-quarter from 350 revolutions of the front roll to 240, and kept the work separate, and it was spun separate, numbered and tested separate, and it was found to be no better than the other yarn made from drawing running at high speed.

Electricity is excited more or less by all railways and drawing frames, and the same causes and cures are applicable to both kinds of frames. The general cause is friction; and when the air is dry it is a non-conductor, and hence the electricity excited remains to trouble the work; and sometimes it acts very badly. To get rid of this annoyance several kinds of treatment are resorted to, all of which are more or less effective. Pointed wires are sometimes suspended over drawing belts, and a channel of escape is thus opened into a sprinkler pipe or some other iron passage; a portion may be gotten rid of in that manner. Steam pipes are sometimes opened, and a little loose steam let out in the room soon dampens the air and makes it a conducter; then the electricity vanishes. But prevention is better than cure. First, have no more weight on rollers than is absolutely necessary; second, keep rollers well oiled—don't let them heat up; third, don't let stirrups chafe the rolls, and avoid all unnecessary friction everywhere; fourth, put a coat of varnish on rolls that are old, rough and dry [See Recipe, Chapter VI.]; but after all has been done as a preventive that can be done, then resort to some of the methods mentioned above to conduct it away. When no more convenient means are at hand, set buckets of hot water about the railways and drawing frames; the air will soon grow moist and take the troublesome fluid, or as it is called in these days, "force," away.

It is essential to good drawing that the top rollers be looked after very closely, both in railways and drawing frames. Where solid rollers are in use, the front ones call for very frequent oiling, owing to the heavy weight necessary. If they get a little warm and dry they will make heavy work, if they don't cut. It won't do to put on much at a time; little and often is the rule; for if the leather gets saturated with oil it soon becomes an old, worthless roller. The patent weighting apparatus saves rolls considerably. Taking the weight off over Sundays and holidays prevents them from being pressed out of shape. This same arrangement is also very useful in removing weights at any time. Owing to the fact that the doffers and feed of cards are driven by the railway and consequently stop when the railway stops, as the main cylinder throws some extra cotton on to the doffers, and when they start are apt to break down more or less ends, thus causing an irregularity in the sheet as it goes to the railway: for these reasons it is desirable to run railways as regularly and stop as seldom as possible and as short a time as may be when they are stopped. In case of a "roll up" it is often necessary to remove the weights from rollers. Railways weighted in the old style take much more time than the new, and on that account, in addition to others mentioned, the new style are preferable to the old.

When any of the small rollers in a railway or drawing frame need to be changed, always work the old ones back, and put the new one in next to the front. Never put a new roll in backside; it can and should be better employed, while an old one will hold very well in that place. Where shell rollers are used, much of the trouble experienced with solid rolls vanishes at once; they are the great invention, so far as top rollers are concerned. Once in two weeks is often enough to oil railway front top shell rollers, rather than once in two hours, as is often the case with solid ones. The writer can hardly refrain from speaking in the highest terms of the shell roller, though it is not the object of this volume to commend any machine; but he may be warranted in doing so, perhaps, under these circumstances: He never was acquainted with the inventor, or any of his agents; never so much as spoke with either; it is not on their account he speaks, although the inventor is most certainly worthy of the highest commendation; but he speaks for the whole cotton manufacturing interest: owners, treasurers, superintendents and all factory operatives: men, women and children. Let it be understood, then, that shell rollers are a saving of oil, rolls and labor; that they make better work, besides cleaner, better looking and more easy to handle.

Modern railways, with all the improvements attached—stop motion, **evener,** patent weighting apparatus and shell roller—are very complete and effective machines. Eveners are worthy the name they bear. When the cotton is dry and sliver needs to be lighter than at any other time, they make it so; because the fibres at such times stand out and create more friction in the trumpet. On the other hand, when the air is damp, and the cotton has absorbed a quantity of water (and should on that account be heavier) the fibres lie down smooth and go through the trumpet easily, and the same bulk is heavier. Again: when an end breaks down, they adapt themselves to the circumstances immediately, and the deficiency is counteracted. Keep them in order, and they are faithful servants.

So much has been said under the head of railways, which are a kind of drawing frame, possessing many of their common characteristics, that as a matter of course what has been said is necessarily equally applicable to both, and but little remains to be said of the latter. Where no railways are used, the first head of drawing may draw more between the back and middle rollers than any that succeed it. [See reason explained under head of railway.] We find drawing frames with three, four and five sets of rolls. With a three or four-roll head the back sets should all be made to hold, and the principal drawing be done between the two front sets, as already explained. With a five roll head they may be, and generally are, arranged as two sets, or more properly do the work of two ordinary frames. The three back sets (two small and one large) correspond to the common three-roll head, and should be managed the same. After passing these it usually passes through a guide before entering the two front sets. During this passage it needs no draught. After entering the two front sets it passes another drawing process, similar to the one already passed: the only difference is one back top roller holds it while being drawn, instead of two, as in the other case. This latter plan is open to objections already made. The front roller of a drawing frame may run anywhere from 350 to 400 revolutions per minute.

There are various devices to fill or crowd cans, such as coilers, plungers and plates, or, as called by some, crowders. The only difficulty with any of the above arrangements is, they are often allowed to run too full and crimp or snarl the drawing. Change the cans when moderately full, and there is no trouble with any of them. The objection to crowders has been so great in some places that they have been entirely dispensed with — no crowders of any kind; the girl crowds the drawing with her hand, and this is generally the most irregular kind of crowding that drawing gets. Drawing is improved by being condensed a little by a groove in the bottom, a tongue in the top calender rolls and a little weight applied to the top roller.

Doubling and drawing are as closely connected as faith and works. If we have one we must have the other; they must go together. If we should undertake to draw card sliver (as delivered from the card without any doubling) enough to lay the fibres parallel, we should get it down so small before the object would be accomplished that we could do nothing with it. But by doubling we are enabled to do this perfectly, and at the same time equalize the different strands so that they are all alike, or nearly so, in weight per yard.

The philosophy of drawing, as applied to cotton, is to displace the fibres, or a portion of them, in such manner as to elongate the strand or thread, as the case may be, and in proportion as it is elongated it is made lighter. This drawing out fibres by rollers running at different speeds is done the easiest and in the most perfect manner when the draught is considerable; the more the draught the fewer fibres are displaced, or drawn out, at a time. The writer would not be understood to intimate that there is no limit to draught. What he wishes to be understood as saying is, that a draught of four can be performed easier and more satisfactory than a draught of two. Just draught enough to separate the fibres is the hardest kind of draught—the hardest for rolls, hardest for the fibres, and it takes more power. This is the reason why it is not well to draw between the two sets of back rollers in any drawing or roving frames. Take up a piece of railway drawing and grip it tightly between the thumb and forefinger of both hands; leave a space of about an inch and a-quarter between the two bites, and then pull hard enough to separate it. That will show the power necessary to draw a little. Then take up a piece of the same kind and hold in one hand as before, and with the other thumb and forefinger take hold of a few fibres and draw them out; repeat rapidly; and that will show the power required to draw considerably, and it will be found to be much less than the former. The draught necessary for back sets of rollers has been explained. The draught of all frames, drawing and roving, should increase as it progresses from one frame to another, with the single exception of the railway-head, or first head of drawing, where there are no railways. There are two reasons for this. First, it is growing lighter; second, it draws easier the farther it goes; that is, it draws easier in a coarse speeder than in the last head of drawing; and it draws easier in an intermediate than in the coarse speeder, and so on.

Top rollers should be perfectly round, straight and both ends of a size. Cover the front ones with good calf skin; sheep skin will do for back ones, though calf is better, more reliable and does not bed down so quick. There is in most modern drawing frames a traverse attached. Rolls wear better with them than without them, though it is necessary to weight a little heavier with the ordinary traverse.

There is a traverse motion in existence, so arranged that the two ends under the same rollers traverse in opposite directions instead of both going one way, as they do on the old plan. It is said they require less weight and are consequently better. The writer has never seen any of them, but has read of them in Draper's book, and judges that they may be an improvement.

Clearers for top rollers in railways and drawing frames should not be allowed to run so long as to have the cloth worn entirely smooth.

The best kind of saddles for solid top rollers in drawing frames are those bushed with green hide, whether hooks or saddles, dead weights or lever weights; and when they get worn, renew them.

It is well to weigh railway drawing twice a day, as a general thing. Have a standard weight per yard, and keep it as near the standard as possible by changing the weight of the evener when it is found to be too light or too heavy.

Drawing frames should be kept clean and tidy and in perfect order, so that they will run well and avoid piecing as much as possible, and also save making waste to be re-worked. The less piecing and less waste there is made the better for the work.

The driving belts should be allowed to run as loosely as they will drive, they stop so much better than with a tight belt, or rather the stop-motion works more satisfactorily and less ends run through.

We may sum up drawing about as follows: The doubling in the railway-head will be according to the number of cards that furnish it, from 6 to 14. Draught, medium, $4\frac{1}{2}$; speed of front roll ($1\frac{1}{2}$ inches diameter), 375 revolutions per minute; first head, double two or three into one draught, $3\frac{1}{4}$; speed of front roll ($1\frac{1}{4}$ inches diameter), 400 revolutions per minute; third head, double three into one, draught $3\frac{1}{2}$ to $3\frac{3}{4}$; speed of front roll ($1\frac{1}{4}$ inches diameter), 400 revolutions per minute.

CHAPTER V.

Speeders.

GENERAL REMARKS — NAMES — TWO KINDS OF POWER — DRAWING WITH ROLLERS — WINDING — TENSION: HOW TO REGULATE — A NEW APPLIANCE — TWIST: HOW TO CALCULATE — HANK: WHAT IS IT — THE OBJECT OF TWIST — A RULE — VARIATIONS WHEN MADE — BALANCE WEIGHTS — SPEED OF ROLLS AND FLYERS IN DIFFERENT FRAMES — FLYERS: HOW BALANCED — RAIL RUNNING BY: HOW TO GUARD AGAINST IT — OIL — BELTS — DANCING BOBBINS: WHAT CAUSES THEM, AND HOW TO CURE — TRAVERSE FOR TRUMPET ROD — RECAPITULATION — PROGRESSIVE DOUBLINGS AND DRAWING.

AFTER the sliver of cotton has been drawn to a certain tenuity, and the number correspondingly multiplied, fibres mostly parallel and pretty well straightened out, in order to continue the process it is necessary to twist it, to make it hold its shape, and also to make it more convenient to handle, as ends increase in number. We used to employ for this purpose, what we called "speeders and stretchers," and though the work is done now in a somewhat similar manner, the number of frames has been increased, plan of construction modified, manner of tending almost a new trade, and names changed. "Speeder and stretcher" have been for many years peculiarly American. While "slubber and fly-frame" have been, and are now, peculiarly English terms. "Speeders" are the names applied to roving frames in this country, within a few years past. Coarse speeders, intermediate speeders and fine speeders compose the list of roving frames; and though there is a striking similarity in general principle between all these frames, yet in detail there is a great difference.

In all these frames there are two motions: one absolute and regular; the other relative and varying. The relative and varying motion or speed is in all cases produced and regulated by the use of a compound and cone or cones. In a speeder the relative power is applied to the spindles and rail, and the absolute to the flyers, while in the slubber and fly frame it is applied to the bobbins on the spindles and to the rail, the absolute to spindles, and to these the flyers are attached. In either case they are about as complex as a watch, and require very skillful management in order to secure good work. Cotton may receive the most approved opening, picking, carding and drawing, and yet it may be all spoiled in the roving frames. Though these frames are so complicated and delicate, they have been brought to a very high state of perfection, and when properly adjusted and cared for will produce most excellent roving.

So far as the drawing with rollers is concerned, it should be regulated the same as drawing frames, as described in a preceding chapter. On a coarse speeder or a slubber it is not customary to double, though it is sometimes done two into one. The draught of a coarse speeder may be from four to four and a half, as required to furnish the right weight.

Next comes twist. [See remarks on twist and twist table in Chapter VI.]

Next comes winding on the bobbin; and there are two or three things in this particular equally applicable to all kinds of roving frames, whether they are the old soft bobbin or the presser, whether they have heads, are straight, or double taper. First, regulate the speed of the rail so that the roving, as it is wound on the barrel of the bobbin, will exactly cover, neither pile up nor show spaces

between the coils, but make a smooth layer—
almost as smooth and straight as the barrel itself;
and as roving varies in size in different seasons of
the year, and in consequence of changes in the atmosphere between dry and damp, it is necessary
to look out for these changes, and make calculations accordingly. There are those who tell us
exactly what the diameter of every variety of
hank roving is, but as they have forgotten to tell
us whether it was in January or in dog-days, when
they measured it, or how hard it was twisted, or
anything of this kind, we can place very little
reliance on such information. The space in this
volume does not permit us to go into calculations
in this direction, and it is a matter that more particularly belongs to machine builders than carders,
though it is well for every carder to understand it;
and the author may at a future time go into such
a work; but let it suffice at present to say, that
machines are usually built with reference to the
hank sought—so near generally that they can be
adjusted with gears to suit the circumstances.
Make the first layer of roving just cover the barrel of the bobbin. Second, never allow any draught
between the front roller and the bobbin; watch
the tension close; keep it loose, for if it draws
hard there it will certainly be spoiled. Drawing
in this place is more injurious than it is in the
drawing frame, between the front fluted and calender rolls, because it is lighter. It is true, we
twist it in a roving frame, but with most kinds of
speeders, and the back row of ends on a fly frame
or slubber, there is a little thin place drawing over
a portion of the rolls so close that the twist does
not run up to the bite of the rolls; and this is the
place where the straining is done. The twist usually runs up better in slubbers and fly frames than

in speeders, for the reason that it don't drop down so suddenly, and the half-turn round of the flyer-nose has a tendency to throw the twist back some. There is a tube put in the nose of the flyer of a speeder which helps throw twist back, and it is a very good thing. A Providence, R. I., firm puts in such tubes when desired. If speeder roll stands inclined a little more from back to front, it would help throw twist back nearer the bite of the rolls. All these things are safeguards; but let it be understood that none of them, nor all of them together, will counteract a draught in this place, and it should never be allowed under any circumstances.

The tension in this place is constantly changing and needs to be watched and changed as occasion calls. On the old soft work, or what we used to call "stretchers and speeders," there is not much chance to change tension conveniently. There is usually a chance to change the cone belt, or rather its position on the cone, at the starting point an inch or more, and that is all; if more is wanted, as is often the case, there must or ought to be a change of racks to finer or coarser as needed. This involves great labor, but it is the only way the writer knows to do it on that kind of frame. There may be some of the old geared cones running yet, somewhere in the country, and if there are it is necessary, in order to change the tension on such, to change an entire set of gears on both cones, and this requires more labor still; but it is the only way to do it. Fly frames and slubbers are easily managed in this particular, and are usually provided with a number of change gears, which are very convenient and may be used to any desired extent. The Lowell speeder double-taper bobbin also has a very convenient change of gears for this purpose.

In addition to what the builders have provided for regulating the tension of these frames, Mr. Jabez Edwards, of the Merrimack Manufacturing Company, Lowell, Mass., has invented and secured a patent on a very useful and convenient appliance to assist in this particular. The guide through which the cone belt runs is so arranged by means of two inclined slots running in opposite directions meeting at the top, and from that point diverging, downward, regulated by a pin passing through the slots and attached to a burr or rod, underneath the regular bar on which the cone belt stand slides. This lower bar is held in a socket in the stand where the spring is coiled, at one end; the other end is held by a bolt in a slide at the end of the cones, and can be lowered or raised at pleasure a distance of some three and a half inches by simply loosening the nut on the bolt. The effect is to change the position of the belt on the cones at once, and also to change the rate of the let-off. Suppose we find a frame drawing or winding too tight, where this patent is attached, we raise the bar a trifle; this throws the cone belt ahead slightly, and the trouble is remedied at once, and at the same time the rate is changed; it now lets off more each time than it did before. Again: suppose we find a frame winding too loose: the girl who tends it would say — "My frame runs slack." We lower the end of the bar a little, and that throws the cone belt back a trifle, and it tightens up immediately, at the same time changing the rate; it lets off less at each time now. The advantage of this appliance is—first, it remedies an existing irregularity at once, which changing gears does not, but they must have considerable time to adapt themselves, or the work, to the new let-off, and it makes some bad work the while; second, it is quickly done; there is no danger of letting the chain slip and the belt on the cone losing its place, the nut being lowered on the builder screw, or anything of the sort; it is all safe, easy, effective and comfortable. This arrangement is equivalent to a change of three teeth in a 50-teeth gear, when it is moved the whole distance; and if more is wanted, the gears are at hand to do it, in addition to the regulator.

The writer does not claim that he has given an intelligent description of this appliance; perhaps it cannot be given without drawings; but he thinks every carder will understand and appreciate its work. Mr. Edwards, the inventor, is a veteran carder, and a man who has invented and applied a great many important improvements to carding machinery. This is the only one he has ever patented, and this one is shut up in a few frames in the Merrimack Manufacturing Company's mills in Lowell, Mass. The reason is his genius, industry and perseverance are only excelled by his modesty. He has scarcely told anybody that he has such a thing; but it is of too much importance to be thus concealed. Cotton manufacturers should have the benefit of it, and they surely will when it is known.

Tension varies more in frames where the presser is used than on soft work, for the reason that the bobbins of roving don't yield so much; but there should be no strain allowed in this place on any kind of frames.

Next in order is twist. It has been a question of considerable importance to carders to know exactly how much to twist roving in order to secure the greatest production, and at the same time make it strong enough to hold and run well, and handle without difficulty. The square root of the hank has been adopted as a basis on which to work—not that it has anything in particular to do with it, only as it forms a regular scale on which to work. There are twist-tables in abundance; they are about as varied as the men who made them. Some twist-table manufacturers understand their business, and others do not; some are made on a regular grading scale, others are not. There are some rules that should be adopted in establishing roving twist for any kind of frames. First, there should be a regular grading scale used; second, the same rule should be used for all kinds, *i. e.*, coarse, intermediate and fine, on the same class of frames.

In order to a proper understanding of this subject, and to begin at the bottom, we will first consider the term "hank." A hank of roving or yarn (cotton) is 840 yards; and when we say half-hank, we mean that 420 yards of such weigh a pound; three-fourths of a hank (630 yards) weigh a pound; one hank (840 yards) weighs a pound; two hanks (1680 yards) weigh a pound; and so on of all kinds. When we use the square root of the hank as a basis to work on in establishing twist, it is the general custom to multiply the square root of the hank by some number uniformly, from the lowest to the highest, as for instance: 1.20 is a favorite multiplier by our English brethren for fly frames. That means to multiply all kinds of hank, from the coarsest slubber to the finest fly frame, by 1.20 to get the proper twist, and the same of any other multiplier. Any one may tell the first time he puts his eye on a twist table (no matter whose) what the maker used for a multiplier, that is, if the table is regular. See what the twist of 1 hank roving is, as the square root of 1 is always 1; whatever it says twist 1 hank, is the multiplier used. We have said that where the square root is used as a basis for calculating twist, it is the general custom to use some number as a multiplier; and it is, though the rule has exceptions, as there are some who use the square root "straight," without any multiplier at all. In that case the twist of 1 hank would be 1 turn of twist to the inch, and so on in that proportion for all kinds of hank. We find different parties in these days using for a multiplier of the square root of the hank as a basis for twist of roving, numbers all the way from 1.05 to 1.48, or rather, we find twist that would require such variety of numbers as that to give them.

The object of twist is to put roving in compact form—so compact as to hold together in good order to wind on the bobbin, as it is made in the frame, and render off in the creels of other roving and spinning frames, without stretching or straining in the process, and that is all the twist that is needed; more is injurious. A twist table will be found in the chapter of tables, with as small intervals as is practicable. The square root of the hank is carefully worked out and placed opposite every number; the fraction is carried out three figures; that is far enough for all practical purposes. The multiplier that is used is 1.24, and it is designed for the Lowell speeder, double taper bobbins. The writer has used such twist on these same frames with success for a number of years. He would recommend that number as a multiplier for the square root of the hank roving for that kind of frames—1.20 for fly frames and slubbers; 1.18 for soft work, where there are no pressers. He would not lay down an arbitrary rule to be followed under all circumstances; but it is well to have a rule, and vary from it when necessary, never unnecessarily nor for any outside reason. It should never enter the mind of a carder to increase or decrease twist, to increase or decrease quantity, though it is often done. Never try to twist roving hard enough to overcome tension in roving frames, bad steps, or dull skewers in roving or spinning frames. After the twist has once been put right it should never be changed for any of the above reasons.

When to vary from the rule: Increase—First, when the roving breaks or stretches between the creel and the back roll; if everything else is right increase twist until it holds; second, unusually high speed of frames sometimes calls for more than ordinary twist; third, unusually heavy bobbins and unusually short staple of cotton need more twist in either case. Decrease—First, when it is noticed that roving worms about between rollers and does not draw well: if everything else is right decrease twist until it draws through straight; second, unusually low speed of frames, light bobbins or long staple may in either case do with less twist.

Steps in creels of roving or spinning frames and the points of skewers should be kept in order all the time, or the roving will suffer. Presses on all kinds of frames should have a uniform power of spring, and be kept at right angles with the bobbin, especially on double-taper bobbins. An intermediate roving frame may draw from 4.75 to 5.10; a fine speeder from 6 to 6.50.

Flyers should be well polished out inside, to insure a smooth passage for roving through them. Balance weights for rails—either bobbin or spindle —should either be lighter than the rail with bobbins empty, or heavier than the rail with bobbins full, as it makes it very bad when they just about balance each other. In that case the rail will run unsteadily and vary in its motion all the back lash will allow, will not wind well, or make a smooth bobbin of roving. They should hang all the time one way or the other. The speed of the front roll in a coarse speeder may be from 180 to 200 revolutions per minute; the flyer, from 675 to 700; intermediate front roll, from 150 to 160; flyer, from 880 to 900; fine front roll, from 112 to 120; flyer, from 1150 to 1200. The improved fly frame and slubber long tubes may run as fast as the speeder, but the old style, where the improvements are wanting, must run more slowly. It is of the greatest importance that flyers are properly balanced; if they are not, in fly frames the spindles will wear one-sided, and in speeders, the noses of flyers wear the same. It is difficult to balance flyers with the presser on, and they must be so balanced. As the presser is constantly changing position, of course the run of the flyer is affected. The proper way to balance them is to do it on the fly and fasten the presser half-way between the barrel of the bobbin and the outside. Balanced in this manner they will throw one way on the start, and the other way when it is full; that is the best that can be done.

It sometimes happens with a slubber or a fly frame, and occasionally with a speeder, that the rail motion does not change and the rail continues to move up or down, as the case may be, until it comes in contact with something strong enough to stop it. When such things take place there is a "smash up," frequently doing a great amount of mischief in a short time. There are various causes that produce such results. In a slubber or fly frame any little stick of the upright shaft, with the bevel gear on top, blanks in it for the driver to run in at all times, except when changing. If this shaft happens to stick, so as not to start when the builder has passed up or down by the arm, then the rail continues to move on, and trouble follows. Of course accidents will sometimes happen, but frequent examinations and careful management will prevent many of them. It is well to have a smart spring drawing at the bottom of this shaft, so that when the builder passes the spring will immediately start and the gear take into it. Sometimes the bevel gears at the top of this shaft do not take in right; then a gear is broken, and perhaps other mischief done. It is quite a trade to set these gears right; it is best to set the first tooth next the blank, about half the thickness of a tooth from the driver, instead of setting it as close as it will run, as is the rule with some. In a speeder or "stretcher," where the rail is carried by cams, the rail never runs by; but the rack and pinion principle as applied to these frames is the same as a fly frame, though the changes are made differently; they sometimes run by. It is important to have everything connected with the change and rail motions adjusted so as to act free. In the Lowell speeder double taper, at the point where the traverse is longest, the rail goes very near the ladder and the floor, so if they do run by they cannot go far, though it makes great trouble when they do. Everything about them should be thoroughly screwed up, to start with, and watched, as wood shrinks and other things take place to loosen nuts.

It is well to have dripping pans under all kinds of roving frames, the whole length and breadth of every kind. This is not generally practised with fly frames, but it would be better to do so, as more or less oil gets on the floor and will in time work through. A mixture of one-third sperm and two-thirds Downie's paraffine oil makes a very good lubricator for carding room machinery—better, in some respects, than clear sperm oil, as it will not gum and is cheaper.

The beauty of a roving frame of any kind is to have it sit level and straight, so that everything will run perfectly easy. If the spindle or bobbin rail goes at all hard, it will strain the roving; so of cones, or the compound, unless the bobbin leads, and in that case the ends will run loose, which is not as bad as straining, though it is bad enough. In either case they should go free, and for the purpose of keeping them so, the suggestions above should be heeded; and in addition, keep them clean and well oiled. Perhaps it should be mentioned that the above-named oil has been charged with staining cotton when it gets on it, and that the stains will not bleach out very readily. All this trouble may be avoided by using clear sperm for rolls, spindles and pressers, and the other for all underneath work.

Cone belts should not be made of hard or stiff leather; if they are it is difficult to make them hold, and if they do not hold well the same results follow that have been mentioned in connection with hard-going rails or spindles. Belts for speeders should not be so tight as to start the frame with a jerk; if they are there is a liability to be a slip of the cone belts and a straining of roving.

Bobbins sometimes "dance" and cause bad winding, and consequently strain roving. There are various causes for dancing bobbins. First, the spindle or bobbin shaft may be out of true and cause a throw. If it is so, straighten it at once. Second, a little bunch of cotton or some other substance may have got into the gears that drive. If so, remove them. Third, gears may be set too close, especially spiral bevels. If that is the case, set them right. Fourth, the bobbin may be loose on the speeder spindle, and the jar of the frame produce the dancing. This is a hard disease to cure. Exposure to a damp air may help such, but don't let them dance.

There should be a traverse to the trumpet rod in all roving frames, and have it work as long as the bosses of the rolls will admit. A very quick traverse is not so good as a more moderate one.

If a roving frame does not let off at any time, and runs up and down two or three times, or until the roving gets strained so much that it begins to break down, and the girl comes and says—"My frame won't run; there don't seem to be any twist in the roving"; then doff that frame, and wind off the roving until it comes good. Never let such pass into the spinning-room; it makes abominable yarn, and will spoil the run of a whole web, perhaps. It is true that this treatment makes waste, and a very bad kind of waste to re-work; but it is the lesser of two evils.

As this is the last process cotton goes through, before spinning, it may be well before closing this chapter to recapitulate a little. A synopsis of the whole thing may be more convenient and satisfactory, in a work of this character, than to be obliged to turn to many different places to find all the particulars.*

First, then, we will begin with the picker. The whole draught in a set of pickers and lappers may vary from 4 to 10; card draught, from 60 to 80;

---

* The author's manuscript is all in the hands of the printer at this time, so far as written and corrected, and he has only two or three sheets on hand; but he thinks he has this subject sufficiently well in mind to go over it very nearly as it has been written, and he trusts that any small variation under these circumstances may be pardoned.

railways, from 4 to 5; first head drawing, from 3 to 3½; second head, from 3¾ to 4; coarse speeder, from 4 to 4½; intermediate, from 4¾ to 5¼; fine frames, 6 to 6½; slubbers and fly frames, same as speeders; speed of beaters in a picker or lapper, ordinarily from 1,400 to 1,800, according to size of beater; card cylinders, from 120 to 160, according to circumstances; front roll in railway, from 325 to 375; drawing, from 350 to 400; coarse speeder, from 180 to 200; intermediate, from 150 to 160; fine frame, from 110 to 120; coarse speeder flyers, from 680 to 700; intermediate, from 850 to 900; fine frame, from 1,000 to 1,100.

These speeds are calculated for modern, improved frames; if they are old-fashioned or shaky, it is not well to run them so fast. For the new Higgins and some other machines the builders claim that their frames, with long bolsters, centrifugal pressers and other improvements, will run at much higher speeds than mentioned above. The writer has had some experience with such, and he would not recommend running any of them at a higher speed than he has mentioned, either for quantity or quality.

One very important matter in carding cotton and preparing it for the spinner is to handle carefully in every process. Don't tear laps, snarl drawing, bang roving about, or let it lie on frames or anywhere else until it is all covered with lint and dust. If it must lie, cover it up. Roving is best when first made; whatever may be the case with yarn, roving does not improve by age.

Progressive doublings and drawing have to be taken into account in making calculations for a new mill or in changing an old one; and though we cannot make exact calculations, always to be depended upon to a fraction of a grain, yet we may approximate very nearly to what we desire by figures; and then, when we come to put work through the machinery, make such changes as are necessary to get the exact thing. The reason why such calculations do not always hold out according to figures, or give such results practically as a correct theory says they will, is owing to the fact that machinery is not perfect in all its parts; that is to say, all the pulleys, rollers and cylinders may not always measure exactly what they are said to measure. Again: some part of the work has to be estimated. It is not known, nor can it be known, exactly how much cotton may be thrown off in the shape of waste in working; but we can usually come near enough to warrant a start, and then, as has been suggested above, make such changes as appear to be necessary to obtain what we desire. Take an example: We will suppose a certain picker lap weighs 10 ounces or 4,375 grains per yard. Whole draught from breaker-lap roll to lap-head delivering roll 67.20, which we use for a divisor for the weight of lap; $4,375 \div 67.20 = 64.53$ grains as the weight of one yard of breaker sliver, as delivered from the card. Next we multiply the weight of one yard by the number of cards doubled into the lap-head, and we will call the number 75: $64.53 \times 75 = 4,839.75$ grains as the weight of one yard of finisher lap; 5 per cent. out for loss in carding where there are no screens, and the flyings re-worked $= 4,594.75$ as the real weight of one yard of finisher-lap ready to be carded. 66.21 draught for finishers, and 4½ per cent. for loss, will make a yard of drawing sliver from finisher-cards weigh 66.27 grains; 1.11 out for draught between card delivery and the back roll of railway$=63.50$ grains as the weight of a yard of finisher drawing, as received by the railway. 63.50×9 (the number of finishers into one railway)=571.50 as the weight of one yard of the sheet, as doubled and enters the railway to the drawing down; 571.50÷4.50 (railway draught)=127 as the weight of one yard of railway drawing, as delivered from that head. 127×2 (doublings in first head of drawing)= 154÷3.12 (draught of first head)=81 grains that a yard of second head drawing should weigh. 81×3 (doubling for third head)=243÷3.57 (draught of third head)=68 grains as the weight of a yard of drawing from the third head. 68÷4.10 (coarse speeder draught)=16.58 grains as the weight of a yard of coarse speeder roving. 16.58×12 (see rule for hank roving)=199 grains to 12 yards. 100.00÷199=$\frac{100}{199}$ hank roving for coarse speeder. 199×2 (doubling for intermediate)=398. 398÷ 5.25 (draught of intermediate)=75.80 grains that 12 yards of intermediate roving will weigh. 100.00÷75.80=1.31 as the hank. 75.80×2= 151.60 doubling for fine frame. 151.60÷6.50 (draught of fine frame)=23.32 as the weight of 12 yards of fine roving. 100.00÷23.32=4.31 hank; and if it is desired it can be taken one step farther and be put into yarn, though this is not carders' business as a general thing. But it is well for every carder to understand all about it.

Suppose the draught of a spinning frame to be 7.25; then 7.25×4.31=31.25 (nearly) as the number of the yarn; but practically we find that it would be heavier than figures show, because the twist in the yarn takes up a little; so we find it necessary to add three per cent. for this shrinkage, or, what is the same thing, subtract three per cent. from the number of the yarn as shown by figures: 31.25×3=.94 (nearly); 31.25−.94=30.31, and that is about the number of yarn that such roving and spinning draught as described will make. The carder can tell what number of yarn the spinner is making at any time if he knows his draught.

The above plan of operation for calculating the numbers for a mill is shown above, in principle; any number desired can be worked out in that manner. Or, reverse the operation; begin with the yarn, and the weight of picker lap necessary to make it will be shown.

The doublings of a mill are reckoned as follows: Suppose three laps are run into one on the finisher picker, as is the case in most modern pickers; if it is only two, then begin with two, or if it is only one and the doublings don't begin until we come to cards, then begin there wherever the doublings commence, but for an example we will call it three into one on the last picker, 75 into one lap-head, 10 into one railway, 2 into first head of drawing, 3 into second head, 2 into intermediate, and 2 into fine frame. They would run thus: $3 \times 75 \times 10 \times 2 \times 3 \times 2 \times 2 = 54,000$, and that is enough for ordinary kinds of work. Doublings are more nowadays than formerly, that is, where the carding is double. Suppose we take an example of single carding to see where we come out, and we will suppose everything is well adapted to that kind of work. We will have to start with the three ends in the picker as before, then the ten ends in the railway and the rest are all the same: $3 \times 10 \times 2 \times 3 \times 2 \times 2 = 720$ doublings; and though it is a fact that so far as evening the work is concerned the drawing and roving frames do that, yet it is another fact that the double carding and more doublings make smoother and better work. The writer once worked in a place where there were 120 doublings made in the following manner: picker 2 into one, single cards 10 into one, drawing 3 into one, fine frames 2 into one: $2 \times 10 \times 3 \times 2 = 120$, and the yarn was not good, and never will be as long as that plan is pursued. Doublings can be carried to any desired extent, but if they are carried too far on drawing it gets very tender and slippery and don't work well.

Perhaps no better idea of the production of a speeder can be given than a table prepared by George Richardson, Esq., Superintendent of the Lowell Machine Shop, Lowell, Mass. He has founded these tables on the actual average production of the Lowell speeder, and he has kindly allowed the author to introduce them here for general information.

Lowell Speeder.—1875.

8¼ in. space.—12 in. traverse.

Tables showing production, &c., under 10-hour system and ordinary circumstances.

No. of Roving.	Grains per yard.	Twist per inch.	Rev. of 1⅜ in. front roll per minute.	Rev. of flyer per minute.	Hanks per day per spindle.	Pounds per day per spindle.
.25	33.33	.640	248.68	625	12.595	50.378
.26	32.05	.653	243.84	625	12.452	47.892
.27	30.86	.665	239.29	625	12.334	45.680
.28	29.76	.677	235.05	625	12.220	43.641
.29	28.73	.689	230.89	625	12.109	41.756
.30	27.77	.701	227.00	625	12.003	40.009
.31	26.88	.713	223.25	625	11.900	38.387
.32	26.04	.725	219.68	625	11.013	36.037
.33	25.25	.735	216.63	625	11.736	35.564
.34	24.51	.746	213.18	625	11.658	34.288
.35	23.81	.758	208.64	625	11.567	33.047
.36	23.15	.768	207.23	625	11.508	31.966
.37	22.52	.778	204.52	925	11.451	30.948
.38	21.93	.789	201.85	625	11.381	29.949
.39	21.36	.799	199.27	625	11.327	29.043
.40	20.83	.809	196.73	625	11.274	28.185
.41	20.32	.819	194.28	625	11.223	27.373
.42	19.84	.829	191.89	625	11.173	26.602
.43	19.38	.840	189.53	625	11.111	25.839
.44	18.94	.849	187.55	625	11.076	25.173
.45	18.52	.859	185.30	625	11.029	24.510
.46	18.12	.869	183.06	625	10.984	23.877
.47	17.73	.878	181.25	625	10.952	23.302
.48	17.36	.887	179.43	625	10.920	22.750
.49	17.00	.896	177.63	625	10.890	22.223
.50	16.66	.905	175.86	625	10.860	21.719
.51	16.34	.914	174.15	625	10.830	21.235
.52	16.02	.923	172.45	625	10.800	20.770
.53	15.72	.932	170.80	625	10.772	20.325
.54	15.43	.941	169.17	625	10.744	19.897
.55	15.15	.950	167.57	625	10.717	19.485
.56	14.88	.957	166.24	625	10.712	19.129
.57	14.62	.966	164.69	625	10.686	18.747
.58	14.37	.975	163.17	625	10.660	18.379
.59	14.12	.983	161.91	625	10.645	18.042
.60	13.88	.992	160.44	625	10.620	17.699

Driving pulleys are 13 inches diameter, 2¾ inches face.
Speed of pulleys=307 revolutions **per minute.**
Cotton on full bobbin=60 **to 64 oz.**
Lengths **of** Speeders, including **pulleys, are as follows:** Those of
20 spindles, 16 ft., 2.5 in. long. | 34 spindles, 26 ft., 1.5 in. long.
22   "   17   "   7.5   " | 36   "   27   "   6.0   "
24   "   19   "   .5   " | 38   "   28   "   11.5   "
26   "   20   "   5.5   " | 40   "   30   "   4.5   "
28   "   21   " 10.5   " | 42   "   31   "   9.5   "
30   "   23   "   3.5   " | 44   "   33   "   2.5   "
32   "   24   "   8.5   " | Width outside=20 inches.

Lowell Speeder.—1875.

6¼ in. space.—9¼ in. traverse.

Tables showing production, &c., under 10-hour system and ordinary circumstances.

No. of Roving.	Grains per yard.	Twist per inch.	Rev. of 1⅜ in. front roll per minute.	Rev. of flyer per minute.	Hanks per day per spindle.	Pounds per day per spindle.
.50	16.66	.905	281.38	900	13.812	27.624
.55	15.15	.950	268.11	900	13.270	24.128
.60	13.88	.990	256.70	900	12.843	21.405
.65	12.82	1.030	246.82	900	12.448	19.151
.70	11.90	1.070	237.67	900	12.083	17.261
.75	11.11	1.108	229.73	900	11.765	15.687
.80	10.42	1.144	222.54	900	11.488	14.361
.85	9.80	1.180	215.77	900	11.229	13.210
.90	9.26	1.215	209.64	900	10.994	12.215
.95	8.77	1.248	204.09	900	10.788	11.356
1.00	8.33	1.280	198.94	900	10.603	10.603
1.05	7.93	1.310	194.09	900	10.442	9.944
1.10	7.58	1.340	189.65	900	10.288	9.353
1.15	7.25	1.370	185.58	900	10.141	8.818
1.20	6.94	1.402	181.68	900	9.986	8.321
1.25	6.66	1.430	177.95	900	9.865	7.892
1.30	6.41	1.459	174.48	900	9.742	7.494
1.35	6.17	1.487	171.20	900	9.631	7.134
1.40	5.95	1.514	168.17	900	9.530	6.807
1.45	5.75	1.540	165.24	900	9.439	6.509
1.50	5.55	1.568	162.40	900	9.339	6.226
1.55	5.38	1.590	159.80	900	9.277	5.985
1.60	5.21	1.619	157.27	900	9.177	5.735
1.65	5.05	1.645	154.82	900	9.097	5.513
1.70	4.90	1.669	152.57	900	9.030	5.312
1.75	4.76	1.693	150.40	900	8.965	5.123

Driving pulleys **are 13** inches diameter, **2¼ inches face.**
Speed of pulleys=442 revolutions per **minute.**
Cotton on full bobbin=25 to 30 oz.
Length **of Speeders,** including pulleys, **are as follows:** Those of

36 spindles, 21 ft.,	8.5 in. long.	50 spindles, 29 ft.,	3.5 in. long.
38 "	22 " 9.5 "	52 "	30 " 4.5 "
40 "	23 " 10.5 "	54 "	31 " 5.5 "
42 "	24 " 11.5 "	56 "	32 " 6.5 "
44 "	26 " .5 "	58 "	33 " 7.5 "
46 "	27 " 1.5 "	60 "	34 " 8.5 "
48 "	28 " 2.5 "	Width outside=20 inches.	

Lowell Speeder.—1875.

5 in. space.—8¼ in. traverse.

Tables showing production, &c., **under 10-hour system** and ordinary circumstances.

No. of Roving.	Grains per yard.	Twist per inch.	Rev. of 1½ in. front roll per minute.	Rev. of flyer per minute.	Hanks per day per spindle.	Pounds per day per spindle.
1.20	6.940	1.402	242.18	1200	12.737	10.614
1.30	6.410	1.459	232.71	1200	12.321	9.478
1.40	5.950	1.514	224.26	1200	11.952	8.537
1.50	5.550	1.568	216.54	1200	11.616	7.744
1.60	5.210	1.619	209.72	1200	11.324	7.077
1.70	4.900	1.669	203.43	1200	11.056	6.503
1.80	4.630	1.718	197.63	1200	10.810	6.005
1.90	4.380	1.764	192.47	1200	10.596	5.579
2.00	4.166	1.805	188.11	1200	10.421	5.210
2.10	3.970	1.855	183.04	1200	10.204	4.859
2.20	3.780	1.898	178.89	1200	10.036	4.562
2.30	3.620	1.940	175.02	1200	9.880	4.295
2.40	3.470	1.983	171.22	1200	9.725	4.052
2.50	3.330	2.022	167.92	1200	9.597	3.839
2.60	3.205	2.063	164.58	1200	9.464	3.640
2.70	3.090	2.103	161.45	1200	9.340	3.459
2.80	2.980	2.141	158.58	1200	9.230	3.296
2.90	2.870	2.180	155.75	1200	9.120	3.145
3.00	2.777	2.214	153.36	1200	9.033	3.011
3.10	2.680	2.253	150.70	1200	8.930	2.881
3.20	2.600	2.290	148.27	1200	8.838	**2.762**
3.30	2.530	2.324	146.10	1200	8.759	2.654
3.40	2.450	2.360	143.87	1200	8.676	2.552
3.50	2.380	2.395	141.77	1200	8.599	2.457
3.60	2.310	2.428	139.84	1200	8.531	2.370
3.70	2.250	2.460	138.02	1200	8.469	2.289
3.80	2.193	2.495	136.09	1200	8.398	2.210
3.90	2.140	2.528	134.31	1200	8.335	2.137
4.00	2.083	2.560	132.63	1200	8.278	2.069
4.10	2.030	2.592	130.99	1200	8.221	2.005
4.20	1.980	2.623	129.44	1200	8.169	1.945
4.30	1.938	2.655	127.88	1200	8.116	1.887
4.40	1.890	2.685	126.46	1200	8.069	1.834
4.50	1.850	2.714	126.26	1200	8.027	1.784
4.60	1.810	2.746	123.65	1200	7.977	1.734
4.70	1.770	2.775	122.35	1200	7.936	1.689
4.80	1.740	2.803	121.13	1200	7.900	1.646
4.90	1.700	2.834	119.80	1200	7.855	1.603
5.00	1.666	2.862	118.64	1200	7.820	1.564

Driving pulleys are 13 inches diameter, 2¼ inches face.
Speed of pulleys=453 revolutions **per** minute.
Cotton on full bobbin=12 to 14 **oz.**
Lengths of Speeders, including pulleys, are as follows: **Those** of 44 spindles, 20 ft., 6 in. long.   64 spindles, 28 ft., 10 in. long.
48   "   22 "   2   "   68   "   30 "   6   "
52   "   23 "   10   "   72   "   32 "   2   "
56   "   25 "   6   "   76   "   33 "   10   "
60   "   27 "   2   "   80   "   35 "   6   "

Width outside=20 **inches.**

It will be noticed by close observers that Mr. Richardson makes calculations to put more twist in roving than does the author of this work; Mr. Richardson uses 1.28 as a multiplier for the square root of the hank.

Great care should be exercised in adjusting the shoe or builder of the Lowell speeder, or any other roving frame where the double taper is used. If the bobbin in filling runs a little more on one taper than the other, on the first taper, the place it runs over the roving will be strained, and the other end will soon grow so slack that the end will break down every time it comes to that place. Mr. Jabez Edwards (who has invented so many useful things) has invented an adjustable shoe, or rather adjustable plates, which are attached to the shoe in such a manner that any irregularities of this kind can be set right at once, without filing or even removing the builder.

The distance of rollers in a roving or in any frame where drawing is done, should be just a trifle more than the longest fibres of cotton; and remember that the bites of rollers are a little closer than the measure from center to center, as the top rollers pressed down cover some space.

It is well to weigh fine roving every day or enough to get the average—say ten bobbins from different parts of the room; keep a record, and on Saturday add and average. It is well to have sufficient of the hank table copied to cover the variations in the same room. Stick it up near the weigher's bench, so that when the average is taken (always weighing twelve yards at a time) reference to the table will show the hank at a glance. Where roving frames have hank clocks, take the number of hanks indicated by the clock, multiply by the number of spindles and divide by the average hank during the week, and it gives the production in pounds. If roving frames have other than hank clocks, it is necessary to ascertain by trial what a set will weigh; then multiply sets by the pounds one set weighs; and don't set it down as fixed forever, that a set will always weigh exactly the same, even though the conditions are apparently the same; try them once in a while, say in the spring and again in the fall; they may vary some. It is well to be regular in weighing, and have the same man do it all the time, if possible. It don't make so much difference where it is done as it does how it is done; that is, it may be done in the railway drawing, fine drawing, coarse, intermediate or fine roving, or all these; but let it be done regularly. Good, nice, sensitive scales are essential to correct weighing and ascertaining hank accurately. Where no better means of testing roving scales can be employed the following may be resorted to: Weigh a good United States nickel five-cent piece; it weighs 77 grains.

CHAPTER VI.

Miscellany.

HANK TABLE: EXPLANATION — RULE TO FIND HANK — TWIST TABLE: EXPLANATION — RULE TO FIND TWIST — HYGROMETER TABLE: EXPLANATION AND USE — TABLE OF QUANTITY OF BELTING IN ROLLS IN CERTAIN CASES: EXPLANATION — PERPETUAL CALENDER: EXPLANATION — ADDRESS TO A COTTON CARDER — RULE TO CALCULATE DRAUGHT — RULE TO CALCULATE SPEED — RECEIPTS FOR MAKING GLUE, CEMENT AND ROLLER VARNISH — WEIGHTS OF VARIOUS ARTICLES — TIME.

HANK TABLE.

Number of yards, 12 in all cases.

Grains.	Hank.	Grains.	Hank.	Grains.	Hank.	Grains.	Hank.	Grains.	Hank.
220	.45	129	.77	91.00	1.09	71.00	1.41	57.75	1.73
214	.46	127	.78	90.00	1.10	70.00	1.42	57.50	1.74
209	.47	126	.79	89.00	1.11	69.50	1.43	57.25	1.75
205	.48	125	.80	88.50	1.12	69.00	1.44	57.00	1.76
202	.49	124	.81	88.00	1.13	68.50	1.45	56.50	1.77
200	.50	123	.82	87.50	1.14	68.00	1.46	56.00	1.78
196	.51	122	.83	87.00	1.15	67.50	1.47	55.50	1.79
193	.52	121	.84	86.00	1.16	67.00	1.48	55.25	1.80
189	.53	120	.85	85.00	1.17	66.50	1.49	55.00	1.81
185	.54	119	.86	84.50	1.18	66.25	1.50	54.75	1.82
182	.55	118	.87	84.00	1.19	66.00	1.51	54.50	1.83
179	.56	117	.88	83.50	1.20	65.50	1.52	54.25	1.84
176	.57	116	.89	83.00	1.21	65.25	1.53	54.00	1.85
173	.58	115	.90	82.00	1.22	65.00	1.54	53.75	1.86
170	.59	113	.91	81.00	1.23	64.50	1.55	53.50	1.87
167	.60	111	.92	80.50	1.24	64.00	1.56	53.25	1.88
164	.61	109	.93	80.00	1.25	63.50	1.57	53.00	1.89
161	.62	107	.94	79.00	1.26	63.00	1.58	52.50	1.90
158	.63	105	.95	78.50	1.27	62.50	1.59	52.25	1.91
155	.64	104	.96	78.00	1.28	62.00	1.60	52.00	1.92
153	.65	103	.97	77.00	1.29	61.50	1.61	51.75	1.93
151	.66	102	.98	76.50	1.30	61.25	1.62	51.50	1.94
149	.67	101	.99	76.00	1.31	61.00	1.63	51.25	1.95
147	.68	100	1.00	75.50	1.32	60.75	1.64	51.00	1.96
145	.69	99	1.01	75.00	1.33	60.50	1.65	50.75	1.97
143	.70	98	1.02	74.50	1.34	60.25	1.66	50.50	1.98
141	.71	97	1.03	74.00	1.35	60.00	1.67	50.25	1.99
139	.72	96	1.04	73.50	1.36	59.50	1.68	50.00	2.00
137	.73	95	1.05	73.00	1.37	59.00	1.69	49.75	2.01
135	.74	94	1.06	72.50	1.38	58.50	1.70	49.50	2.02
133	.75	93	1.07	72.00	1.39	58.25	1.71	49.25	2.03
131	.76	92	1.08	71.50	1.40	58.00	1.72	49.00	2.04

HANK TABLE—Continued.

Grains.	Hank.	Grains.	Hank.	Grains.	Hank.	Grains.	Hank.	Grains.	Hank.
48.75	2.05	38.60	2.59	31.90	3.13	27.30	3.67	23.75	4.21
48.50	2.06	38.50	2.60	31.80	3.14	27.20	3.68	23.70	4.22
48.25	2.07	38.40	2.61	31.70	3.15	27.10	3.69	23.65	4.23
48.00	2.08	38.20	2.62	31.60	3.16	27.00	3.70	23.60	4.24
47.75	2.09	38.00	2.63	31.50	3.17	26.90	3.71	23.50	4.25
47.50	2.10	37.85	2.64	31.40	3.18	26.80	3.72	23.45	4.26
47.25	2.11	37.70	2.65	31.30	3.19	26.75	3.73	23.40	4.27
47.00	2.12	37.55	2.66	31.20	3.20	26.70	3.74	23.35	4.28
46.75	2.13	37.40	2.67	31.10	3.21	26.60	3.75	23.30	4.29
46.50	2.14	37.25	2.68	31.00	3.22	26.55	3.76	23.25	4.30
46.25	2.15	37.12	2.69	30.90	3.23	26.50	3.77	23.20	4.31
46.00	2.16	37.00	2.70	30.80	3.24	26.45	3.78	23.10	4.32
45.80	2.17	36.85	2.71	30.70	3.25	26.40	3.79	23.05	4.33
45.60	2.18	36.70	2.72	30.60	3.26	26.30	3.80	23.00	4.34
45.40	2.19	36.55	2.73	30.50	3.27	26.25	3.81	22.95	4.35
45.20	2.20	36.40	2.74	30.40	3.28	26.20	3.82	22.90	4.36
45.00	2.21	36.25	2.75	30.30	3.29	26.15	3.83	22.85	4.37
44.80	2.22	36.12	2.76	30.20	3.30	26.10	3.84	22.80	4.38
44.60	2.23	36.00	2.77	30.10	3.31	26.00	3.85	22.75	4.39
44.40	2.24	35.90	2.78	30.00	3.32	25.90	3.86	22.70	4.40
44.20	2.25	35.80	2.79	29.90	3.33	25.80	3.87	22.65	4.41
44.00	2.26	35.60	2.80	29.80	3.34	25.75	3.88	22.60	4.42
43.80	2.27	35.50	2.81	29.70	3.35	25.70	3.89	22.55	4.43
43.60	2.28	35.40	2.82	29.60	3.36	25.65	3.90	22.50	4.44
43.40	2.29	35.30	2.83	29.55	3.37	25.60	3.91	22.45	4.45
43.30	2.30	35.20	2.84	29.50	3.38	25.50	3.92	22.40	4.46
43.20	2.31	35.00	2.85	29.45	3.39	25.40	3.93	22.35	4.47
43.10	2.32	34.90	2.86	29.40	3.40	25.30	3.94	22.30	4.48
43.00	2.33	34.80	2.87	29.30	3.41	25.25	3.95	22.25	4.49
42.80	2.34	34.60	2.88	29.20	3.42	25.20	3.96	22.20	4.50
42.60	2.35	34.50	2.89	29.10	3.43	25.15	3.97	22.15	4.51
42.40	2.36	34.30	2.90	29.00	3.44	25.10	3.98	22.10	4.52
42.20	2.37	34.20	2.91	28.95	3.45	25.05	3.99	22.05	4.53
42.00	2.38	34.10	2.92	28.90	3.46	25.00	4.00	22.00	4.54
41.80	2.39	33.90	2.93	28.80	3.47	24.90	4.01	21.95	4.55
41.60	2.40	33.80	2.94	28.70	3.48	24.85	4.02	21.90	4.56
41.40	2.41	33.70	2.95	28.60	3.49	24.80	4.03	21.85	4.57
41.20	2.42	33.60	2.96	28.50	3.50	24.75	4.04	21.80	4.58
41.00	2.43	33.50	2.97	28.40	3.51	24.70	4.05	21.75	4.59
40.85	2.44	33.40	2.98	28.30	3.52	24.65	4.06	21.70	4.60
40.70	2.45	33.30	2.99	28.25	3.53	24.60	4.07	21.65	4.61
40.55	2.46	33.20	3.00	28.20	3.54	24.50	4.08	21.60	4.62
40.40	2.47	33.10	3.01	28.10	3.55	24.40	4.09	21.55	4.63
40.25	2.48	33.00	3.02	28.05	3.56	24.35	4.10	21.50	4.65
40.12	2.49	32.95	3.03	28.00	3.57	24.30	4.11	21.45	4.66
40.00	2.50	32.90	3.04	27.90	3.58	24.25	4.12	21.40	4.67
39.85	2.51	32.80	3.05	27.80	3.59	24.20	4.13	21.35	4.68
39.70	2.52	32.75	3.06	27.75	3.60	24.15	4.14	21.30	4.69
39.55	2.53	32.70	3.07	27.70	3.61	24.10	4 15	21.25	4.70
39.40	2.54	32.60	3.08	27.60	3.62	24.00	4.16	21.20	4.71
39.25	2.55	32.55	3.09	27.50	3.63	23.95	4.17	21.15	4.72
39.12	2.56	32.40	3.10	27.45	3.64	23.90	4.18	21.10	4.73
39.00	2.57	32.20	3.11	27.40	3.65	23.85	4.19	21.05	4.75
38.80	2.58	32.00	3.12	27.35	3.66	23.80	4.20	21.00	4.76

HANK TABLE—Continued.

Grains.	Hank.	Grains.	Hank.	Grains.	Hank.	Grains.	Hank.	Grains.	Hank.
20.95	4.77	18.87	5.30	17.24	5.79	15.92	6.28	14.75	6.78
20.90	4.78	18.85	5.31	17.22	5.80	15.90	6.29	14.73	6.79
20.85	4.79	18.81	5.32	17.20	5.81	15.87	6.30	14.70	6.80
20.80	4.80	18.77	5.33	17.17	5.82	15.84	6.31	14.67	6.82
20.75	4.82	18.73	5.34	17.14	5.83	15.82	6.32	14.64	6.84
20.70	4.83	18.69	5.35	17.12	5.84	15.80	6.33	14.60	6.86
20.65	4.84	18.65	5.36	17.09	5.85	15.77	6.34	14.56	6.88
20.60	4.85	18.61	5.37	17.06	5.86	15.74	6.35	14.50	6.90
20.55	4.86	18.58	5.38	17.03	5.87	15.71	6.36	14.45	6.92
20.50	4.87	18.53	5.39	17.00	5.88	15.69	6.37	14.40	6.94
20.45	4.88	18.49	5.40	16.97	5.89	15.66	6.38	14.36	6.96
20.40	4.89	18.45	5.41	16.95	5.90	15.64	6.39	14.32	6.98
20.35	4.91	18.42	5.42	16.92	5.91	15.62	6.40	14.28	7.00
20.30	4.92	18.39	5.43	16.90	5.92	15.60	6.41	14.24	7.02
20.25	4.94	18.36	5.44	16.87	5.93	15.57	6.42	14.20	7.04
20.20	4.95	18.33	5.45	16.85	5.94	15.54	6.43	14.16	7.06
20.15	4.97	18.29	5.46	16.82	5.95	15.52	6.44	14.12	7.08
20.10	4.98	18.26	5.47	16.80	5.96	15.50	6.45	14.08	7.10
20.05	4.99	18.23	5.48	16.77	5.97	15.47	6.46	14.04	7.12
20.00	5.00	18.20	5.49	16.74	5.98	15.44	6.47	14.00	7.14
19.95	5.01	18.17	5.50	16.71	5.99	15.42	6.48	12.50	8.00
19.90	5.02	18.14	5.51	16.68	6.00	15.40	6.49	11.00	9.00
19.85	5.03	18.11	5.52	16.65	6.01	15.37	6.50	10.00	10.00
19.83	5.04	18.08	5.53	16.62	6.02	15.35	6.51	9.00	11.00
19.80	5.05	18.04	5.54	16.59	6.03	15.32	6.52	8.33	12.00
19.76	5.06	18.00	5.55	16.56	6.04	15.30	6.53	7.69	13.00
19.71	5.07	17.96	5.56	16.53	6.05	15.27	6.54	7.12	14.00
19.67	5.08	17.92	5.57	16.50	6.06	15.25	6.55	6.66	15.00
19.63	5.09	17.89	5.58	16.48	6.07	15.23	6.56	6.25	16.00
19.60	5.10	17.85	5.59	16.45	6.08	15.20	6.57	5.90	17.00
19.57	5.11	17.82	5.60	16.42	6.09	15.18	6.58	5.55	18.00
19.52	5.12	17.78	5.61	16.39	6.10	15.16	6.59	5.26	19.00
19.48	5.13	17.75	5.62	16.36	6.11	15.14	6.60	5.00	20.00
19.44	5.14	17.72	5.63	16.33	6.12	15.11	6.61	4.76	21.00
19.40	5.15	17.69	5.64	16.30	6.13	15.09	6.62	4.55	22.00
19.36	5.16	17.66	5.65	16.27	6.14	15.07	6.63	4.34	23.00
19.32	5.17	17.63	5.66	16.25	6.15	15.05	6.64	4.16	24.00
19.28	5.18	17.60	5.67	16.22	6.16	15.03	6.65	4.00	25.00
19.24	5.19	17.57	5.68	16.20	6.17	15.00	6.66	3.85	26.00
19.20	5.20	17.53	5.69	16.17	6.18	14.97	6.67	3.70	27.00
19.17	5.21	17.50	5.70	16.14	6.19	14.94	6.68	3.57	28.00
19.13	5.22	17.47	5.71	16.12	6.20	14.92	6.69	3.45	29.00
19.09	5.23	17.44	5.72	16.09	6.21	14.90	6.70	3.33	30.00
19.05	5.24	17.41	5.73	16.06	6.22	14.88	6.71	3.22	31.00
19.02	5.25	17.38	5.74	16.04	6.23	14.86	6.72	3.12	32.00
19.00	5.26	17.35	5.75	16.02	6.24	14.85	6.73	3.03	33.00
18.97	5.27	17.32	5.76	16.00	6.25	14.82	6.74		
18.94	5.28	17.29	5.77	15.97	6.26	14.80	6.75		
18.91	5.29	17.26	5.78	15.94	6.27	14.77	6.76		

EXPLANATION OF HANK TABLE.

We have already seen that a hank is 840 yards of roving. To ascertain what hank any particular kind of roving is—that is, how many hanks there are to the pound—it is necessary to weigh a certain number of yards. Grains are generally used in weighing roving, but the pounds (that is, how many hanks to the pound) are avoirdupois; so in making this calculation we must reduce troy-weight to avoirdupois. We find that in one pound avoirdupois there are 7,000 grains; so if we should reel off 840 yards, and it should weigh 7,000 grains, that would be one-hank roving; 3,500 would be two-hank; 1,750 four-hank, and so on; but this would be a tedious job, and so we take a certain part of 840 yards and a corresponding part of 7,000 grains and use the parts instead of the whole. 12 yards are $\frac{1}{70}$ of 840; 100 is $\frac{1}{70}$ of 7,000; so we weigh 12 yards and divide 100 by the number of grains' weight, and that gives us the hank.

EXAMPLES.

12 yards weigh 25 grains:

$$25)100(4. \quad \text{The hank is 4.}$$
$$\underline{100}$$

Again: 12 yards weigh 25.50:

$$25.50)100.00(3.92. \quad \text{The hank is } 3.92.$$
$$\underline{7650}$$
$$23500$$
$$\underline{22950}$$
$$5500$$
$$\underline{5100}$$
$$400$$

Once more : 12 yards weigh 24.50 :

$$24.50 \overline{)100.00}(4.08.\quad \text{The hank is } 4.08.$$
$$\underline{9800}$$
$$20000$$
$$\underline{19600}$$
$$400$$

It is most convenient to put fractions in form of decimals as in the above examples. If 12 yards, or the average of several weighings, weigh a certain number of grains and a fraction, as is most always the case, then add cyphers to the dividend to balance the divisor.

From the foregoing examples and explanations we derive the following rule to find the hank :—First, weigh 12 yards of any kind of roving; second, take the number of grains 12 yards weigh as the divisor for 100; the quotient will be the hank sought.

There are a number of rules to find hank; the writer has some half-dozen, more or less; but the above is his best, so he gives no other.

TWIST TABLE.

Hank.	Sq. Root.	Twist.	Hank.	Sq. Root.	Twist.	Hank.	Sq. Root.	Twist.
.50	.7071	.88	1.03	1.014	1.26	1.56	1.248	1.55
.51	.7141	.89	1.04	1.019	1.26	1.57	1.252	1.55
.52	.7211	.89	1.05	1.024	1.27	1.58	1.256	1.56
.53	.7280	.90	1.06	1.029	1.28	1.59	1.260	1.56
.54	.7348	.91	1.07	1.034	1.28	1.60	1.264	1.57
.55	.7416	.92	1.08	1.039	1.29	1.61	1.268	1.57
.56	.7483	.93	1.09	1.044	1.29	1.62	1.272	1.57
.57	.7549	.94	1.10	1.048	1.30	1.63	1.276	1.58
.58	.7615	.95	1.11	1.053	1.31	1.64	1.280	1.58
.59	.7681	.95	1.12	1.058	1.31	1.65	1.284	1.59
.60	.7745	.96	1.13	1.063	1.32	1.66	1.288	1.59
.61	.7810	.97	1.14	1.067	1.32	1.67	1.292	1.60
.62	.7874	.98	1.15	1.072	1.33	1.68	1.296	1.60
.63	.7937	.98	1.16	1.077	1.34	1.69	1.300	1.61
.64	.8000	.99	1.17	1.081	1.34	1.70	1.303	1.61
.65	.8062	1.00	1.18	1.086	1.35	1.71	1.307	1.62
.66	.8124	1.00	1.19	1.090	1.35	1.72	1.311	1.62
.67	.8185	1.01	1.20	1.095	1.36	1.73	1.315	1.63
.68	.8246	1.02	1.21	1.100	1.36	1.74	1.319	1.63
.69	.8306	1.03	1.22	1.104	1.37	1.75	1.322	1.64
.70	.8366	1.04	1.23	1.109	1.38	1.76	1.326	1.64
.71	.8426	1.04	1.24	1.113	1.38	1.77	1.330	1.65
.72	.8485	1.05	1.25	1.118	1.39	1.78	1.334	1.65
.73	.8544	1.06	1.26	1.122	1.39	1.79	1.337	1.65
.74	.8602	1.07	1.27	1.126	1.40	1.80	1.341	1.66
.75	.8660	1.07	1.28	1.131	1.40	1.81	1.345	1.66
.76	.8717	1.08	1.29	1.135	1.41	1.82	1.349	1.67
.77	.8774	1.09	1.30	1.140	1.41	1.83	1.352	1.67
.78	.8831	1.10	1.31	1.144	1.42	1.84	1.356	1.68
.79	.8888	1.10	1.32	1.148	1.42	1.85	1.360	1.68
.80	.8944	1.11	1.33	1.153	1.43	1.86	1.363	1.69
.81	.9000	1.12	1.34	1.157	1.43	1.87	1.367	1.69
.82	.9055	1.12	1.35	1.161	1.44	1.88	1.371	1.70
.83	.9110	1.13	1.36	1.166	1.44	1.89	1.374	1.70
.84	.9165	1.14	1.37	1.170	1.45	1.90	1.378	1.71
.85	.9219	1.14	1.38	1.174	1.45	1.91	1.382	1.71
.86	.9273	1.15	1.39	1.178	1.46	1.92	1.385	1.71
.87	.9327	1.16	1.40	1.183	1.46	1.93	1.389	1.72
.88	.9380	1.16	1.41	1.187	1.47	1.94	1.392	1.72
.89	.9433	1.17	1.42	1.191	1.47	1.95	1.396	1.73
.90	.9486	1.18	1.43	1.195	1.48	1.96	1.400	1.73
.91	.9539	1.18	1.44	1.200	1.48	1.97	1.403	1.74
.92	.9591	1.19	1.45	1.204	1.49	1.98	1.407	1.74
.93	.9643	1.20	1.46	1.208	1.49	1.99	1.410	1.75
.94	.9695	1.20	1.47	1.212	1.50	2.00	1.414	1.75
.95	.9746	1.21	1.48	1.216	1.50	2.01	1.417	1.76
.96	.9797	1.21	1.49	1.220	1.51	2.02	1.421	1.76
.97	.9848	1.22	1.50	1.224	1.51	2.03	1.424	1.76
.98	.9899	1.22	1.51	1.228	1.52	2.04	1.428	1.77
.99	.9949	1.23	1.52	1.232	1.53	2.05	1.431	1.77
1.00	1.000	1.24	1.53	1.236	1.54	2.06	1.435	1.78
1.01	1.004	1.24	1.54	1.240	1.54	2.07	1.438	1.78
1.02	1.009	1.25	1.55	1.244	1.54	2.08	1.442	1.79

TWIST TABLE—Continued.

Hank.	Sq. Root.	Twist.	Hank.	Sq. Root.	Twist.	Hank.	Sq. Root.	Twist.
2.09	1.445	1.79	3.06	1.749	2.17	4.69	2.165	2.68
2.10	1.449	1.80	3.09	1.757	2.18	4.72	2.172	2.69
2.11	1.452	1.80	3.12	1.766	2.19	4.75	2.179	2.70
2.12	1.456	1.80	3.15	1.774	2.20	4.78	2.186	2.71
2.13	1.459	1.81	3.18	1.783	2.21	4.81	2.193	2.72
2.14	1.462	1.81	3.21	1.791	2.22	4.84	2.200	2.73
2.15	1.466	1.82	3.24	1.800	2.23	4.87	2.206	2.74
2.16	1.469	1.82	3.27	1.808	2.24	4.90	2.213	2.74
2.17	1.473	1.83	3.30	1.816	2.25	4.94	2.222	2.75
2.18	1.476	1.83	3.33	1.824	2.26	4.97	2.229	2.76
2.19	1.479	1.83	3.36	1.833	2.27	5.00	2.236	2.77
2.20	1.483	1.84	3.39	1.841	2.28	5.04	2.244	2.78
2.22	1.489	1.85	3.42	1.849	2.29	5.08	2.253	2.79
2.24	1.496	1.85	3.45	1.857	2.30	5.12	2.262	2.80
2.26	1.503	1.86	3.48	1.865	2.31	5.16	2.271	2.81
2.28	1.509	1.87	3.51	1.873	2.32	5.20	2.280	2.82
2.30	1.516	1.88	3.54	1.881	2.33	5.24	2.286	2.83
2.32	1.523	1.89	3.57	1.889	2.34	5.28	2.297	2.84
2.34	1.529	1.90	3.60	1.897	2.35	5.32	2.306	2.85
2.36	1.536	1.90	3.63	1.905	2.36	5.36	2.315	2.86
2.38	1.542	1.91	3.66	1.913	2.37	5.40	2.323	2.87
2.40	1.549	1.92	3.69	1.920	2.38	5.44	2.332	2.88
2.42	1.555	1.93	3.72	1.928	2.39	5.48	2.340	2.89
2.44	1.562	1.94	3.75	1.936	2.40	5.52	2.349	2.90
2.46	1.568	1.95	3.78	1.944	2.41	5.56	2.357	2.91
2.48	1.574	1.95	3.81	1.951	2.42	5.59	2.364	2.93
2.50	1.581	1.96	3.84	1.959	2.43	5.62	2.370	2.94
2.52	1.587	1.97	3.87	1.967	2.44	5.66	2.379	2.95
2.54	1.593	1.98	3.90	1.974	2.45	5.70	2.387	2.96
2.56	1.600	1.99	3.93	1.982	2.46	5.74	2.395	2.97
2.58	1.606	1.99	3.96	1.989	2.47	5.78	2.404	2.98
2.60	1.612	2.00	3.99	1.997	2.48	5.82	2.412	2.99
2.62	1.618	2.01	4.02	2.004	2.49	5.86	2.420	3.00
2.64	1.624	2.02	4.05	2.012	2.50	5.90	2.428	3.01
2.66	1.630	2.03	4.08	2.019	2.51	5.94	2.437	3.02
2.68	1.637	2.03	4.11	2.027	2.51	5.98	2.445	3.03
2.70	1.643	2.04	4.14	2.034	2.52	6.02	2.453	3.04
2.72	1.649	2.04	4.17	2.042	2.53	6.06	2.461	3.05
2.74	1.655	2.05	4.20	2.049	2.54	6.10	2.469	3.06
2.76	1.661	2.05	4.23	2.056	2.55	6.14	2.477	3.07
2.78	1.667	2.06	4.26	2.063	2.56	6.18	2.485	3.08
2.80	1.673	2.07	4.29	2.071	2.57	6.22	2.493	3.09
2.82	1.679	2.08	4.32	2.078	2.58	6.26	2.501	3.10
2.84	1.685	2.09	4.35	2.085	2.59	6.30	2.509	3.11
2.86	1.691	2.10	4.38	2.092	2.60	6.34	2.517	3.12
2.88	1.697	2.10	4.41	2.100	2.60	6.38	2.525	3.13
2.90	1.702	2.11	4.44	2.107	2.61	6.42	2.533	3.14
2.92	1.708	2.12	4.47	2.114	2.62	6.46	2.541	3.15
2.94	1.714	2.13	4.50	2.121	2.63	6.50	2.549	3.16
2.96	1.720	2.14	4.54	2.130	2.64	6.54	2.557	3.17
2.98	1.726	2.14	4.58	2.140	2.65	6.58	2.565	3.18
3.00	1.732	2.15	4.62	2.149	2.66	6.62	2.572	3.19
3.03	1.740	2.16	4.66	2.158	2.67	6.66	2.580	3.20

11*

TWIST TABLE—Continued.

Hank.	Sq. Root.	Twist.	Hank.	Sq. Root.	Twist.	Hank.	Sq. Root.	Twist.
6.70	2.588	3.21	9.05	3.008	3.73	11.73	3.424	4.25
6.74	2.596	3.22	9.10	3.016	3.74	11.78	3.432	4.26
6.78	2.603	3.23	9.15	3.024	3.75	11.83	3.439	4.27
6.82	2.611	3.24	9.20	3.033	3.76	11.89	3.448	4.28
6.86	2.619	3.25	9.25	3.041	3.77	11.95	3.456	4.29
6.92	2.630	3.26	9.30	3.049	3.78	12.00	3.464	4.30
6.96	2.638	3.27	9.35	3.057	3.79	12.06	3.472	4.31
7.00	2.645	3.28	9.40	3.065	3.80	12.12	3.481	4.32
7.04	2.653	3.29	9.45	3.074	3.81	12.18	3.489	4.33
7.08	2.660	3.30	9.50	3.082	3.82	12.24	3.498	4.34
7.12	2.668	3.31	9.55	3.090	3.83	12.30	3.507	4.35
7.16	2.675	3.32	9.60	3.098	3.84	12.36	3.515	4.36
7.20	2.683	3.33	9.65	3.106	3.85	12.42	3.524	4.37
7.24	2.690	3.34	9.70	3.114	3.86	12.48	3.532	4.38
7.28	2.695	3.35	9.75	3.122	3.87	12.54	3.541	4.39
7.32	2.705	3.35	9.80	3.130	3.88	12.60	3 549	4.40
7.36	2.712	3.36	9.85	3.138	3.89	12.66	3 558	4.41
7.40	2.720	3.37	9.90	3.146	3.90	12.72	3.566	4.42
7.44	2.727	3.38	9.95	3.154	3.91	12.78	3.574	4.43
7.48	2.734	3.39	10.00	3.162	3.92	12.84	3.583	4.44
7.52	2.742	3.40	10.05	3.170	3.93	12.90	3.591	**4.45**
7.56	2.749	3.41	10.10	3.178	3.94	12.96	3.600	4.46
7.60	2.756	3.42	10.15	3.185	3.95	13.02	3.608	4.47
7.65	2.765	3.43	10.20	3.193	3.96	13.08	3.616	4.48
7.70	2.774	3.44	10.25	3.201	3.97	13.14	3.624	4.49
7.75	2.783	3.45	10.30	3.209	3.98	13.20	3 633	4.50
7.80	2.792	3.46	10.35	3.217	3.99	13.26	3.641	4.51
7.85	2.801	3.47	10.40	3.224	4.00	13.32	3.649	4.52
7.90	2.810	3.48	10.45	3.232	4.01	13.38	3.657	4.53
7.95	2.819	3.49	10.50	3.240	4.02	13.45	3.667	4.54
8.00	2.828	3.50	10.55	3.248	4.03	13.50	3.674	4.55
8 04	2.835	3.51	10.60	3.255	4.04	13.55	3.681	4.56
8.08	2.842	3.52	10.65	3.263	4.05	13.60	3.687	4.57
8.12	2.849	3.53	10.70	3.271	4.06	13.65	3.694	4.58
8.16	2.856	3.54	10.75	3.278	4.07	13.71	3.702	4.59
8.20	2.863	3 55	10.80	3.286	4.08	13.76	3 709	4.60
8.25	2.872	3.56	10.85	3.293	4.09	13.82	3.717	4.61
8.30	2.880	3.57	10.90	3.301	4.10	13.88	3.725	4.62
8.35	2.889	3.58	10.95	3.309	4.11	13.94	3.733	4.63
8.40	2.898	3.59	11.00	3.316	4.11	14.00	3.741	4.64
8.45	2.906	3.60	11.06	3.325	4.12	14.06	3.749	4.65
8.50	2.915	3.61	11.12	3.334	4.13	14.12	3.757	4.66
8.54	2.922	3.62	11.18	3.343	4.14	14.18	3.765	4.67
8.58	2 929	3.63	11.23	3.351	4.15	14.24	3.773	4.68
8.62	2.935	3.64	11.28	3.358	4.16	14.30	3.781	4.69
8.66	2.942	3.65	11.33	3.366	4.17	14.36	3.789	4.70
8.70	2.949	3.66	11.38	3.373	4.18	14.42	3.797	4.71
8.75	2.958	3.67	11.43	3.380	4.19	14.48	3.805	4.72
8.80	2.966	3.68	11.48	3 388	4.20	14.54	3.813	4 73
8.85	2.974	3.69	11.53	3.395	4.21	14.60	3.820	4.74
8.90	2.982	3.70	11.58	3.402	4.22	14.66	3.828	4.75
8.95	2.901	3.71	11.63	3.410	4.23	14.72	3.836	4.76
9.00	3.000	3.72	11.68	3.417	4.24	14.78	3.844	4.77

TWIST TABLE—Continued.

Hank.	Sq. Root.	Twist.	Hank.	Sq. Root.	Twist.	Hank.	Sq. Root.	Twist.
14.84	3.852	4.78	18.38	4.287	5.31	22.24	4.715	5.84
14.90	3.860	4.79	18.45	4.295	5.32	22.32	4.724	5.85
14.96	3.867	4.80	18.52	4.303	5.33	22.40	4.732	5.86
15.02	3.875	4.81	18.59	4.311	5.34	22.48	4.741	5.87
15.09	3.884	4.82	18.66	4.319	5.35	22.55	4.748	5.88
15.16	3.893	4.83	18.73	4.327	5.36	22.62	4.756	5.89
15.23	3.902	4.84	18.80	4.335	5.37	22.68	4.762	5.90
15.30	3.911	4.85	18.87	4.343	5.38	22.75	4.769	5.91
15.37	3.920	4.86	18.94	4.352	5.39	22.82	4.777	5.92
15.44	3.929	4.87	19.00	4.358	5.40	22.89	4.784	5.93
15.51	3.938	4.88	19.07	4.366	5.41	22.96	4.791	5.94
15.58	3.947	4.89	19.14	4.374	5.42	23.04	4.800	5.95
15.65	3.956	4.90	19.21	4.382	5.43	23.10	4.806	5.96
15.72	3.964	4.91	19.28	4.390	5.44	23.18	4.814	5.97
15.78	3.972	4.92	19.35	4.399	5.45	23.26	4.822	5.98
15.84	3.979	4.93	19.42	4.406	5.46	23.35	4.832	5.99
15.90	3.987	4.94	19.49	4.414	5.47	23.44	4.841	6.00
15.96	3.994	4.95	19.56	4.422	5.48	23.52	4.847	6.01
16.00	4.000	4.96	19.63	4.430	5.49	23.61	4.859	6.02
16.07	4.008	4.97	19.70	4.438	5.50	23.68	4.866	6.03
16.13	4.016	4.98	19.77	4.446	5.51	23.76	4.874	6.04
16.20	4.024	4.99	19.84	4.454	5.52	23.84	4.882	6.05
16.27	4.033	5.00	19.91	4.462	5.53	23.92	4.890	6.06
16.34	4.042	5.01	20.00	4.472	5.54	24.00	4.898	6.07
16.42	4.052	5.02	20.10	4.483	5.55	24.08	4.907	6.08
16.50	4.062	5.03	20.17	4.491	5.56	24.16	4.915	6.09
16.57	4.070	5.04	20.21	4.495	5.57	24.24	4.923	6.10
16.64	4.079	5.05	20.25	4.500	5.58	24.32	4.931	6.11
16.70	4.086	5.06	20.32	4.507	5.59	24.40	4.939	6.12
16.75	4.092	5.07	20.40	4.516	5.60	24.48	4.943	6.13
16.81	4.100	5.08	20.48	4.525	5.61	24.57	4.957	6.14
16.88	4.108	5.09	20.56	4.534	5.62	24.65	4.965	6.15
16.94	4.115	5.10	20.64	4.543	5.63	24.73	4.975	6.16
17.00	4.123	5.11	20.75	4.555	5.64	24.80	4.979	6.17
17.07	4.131	5.12	20.82	4.562	5.65	24.87	4.986	6.18
17.14	4.140	5.13	20.90	4.571	5.66	24.94	4.993	6.19
17.21	4.148	5.14	20.98	4.580	5.67	25.00	5.000	6.20
17.28	4.156	5.15	21.05	4.588	5.68	25.10	5.009	6.21
17.35	4.165	5.16	21.12	4.595	5.69	25.20	5.019	6.22
17.42	4.173	5.17	21.20	4.604	5.70	25.27	5.026	6.23
17.49	4.182	5.18	21.25	4.610	5.71	25.34	5.033	6.24
17.56	4.190	5.19	21.32	4.617	5.72	25.42	5.041	6.25
17.63	4.198	5.20	21.40	4.626	5.73	25.50	5.049	6.26
17.70	4.207	5.21	21.47	4.633	5.74	25.60	5.059	6.27
17.77	4.215	5.22	21.55	4.642	5.75	25.68	5.067	6.28
17.84	4.223	5.23	21.62	4.649	5.76	25.75	5.074	6.29
17.91	4.232	5.24	21.70	4.658	5.77	25.82	5.081	6.30
17.97	4.239	5.25	21.77	4.665	5.78	25.91	5.090	6.31
18.03	4.246	5.26	21.84	4.673	5.79	26.00	5.099	6.32
18.10	4.254	5.27	21.90	4.679	5.80	26.10	5.108	6.33
18.17	4.262	5.28	22.00	4.690	5.81	26.18	5.116	6.34
18.24	4.270	5.29	22.08	4.698	5.82	26.25	5.123	6.35
18.31	4.279	5.30	22.16	4.707	5.83	26.33	5.131	6.36

TWIST TABLE—Continued.

Hank.	Sq. Root.	Twist.	Hank.	Sq. Root.	Twist.	Hank.	Sq. Root.	Twist.
26.40	5.138	6.37	28.70	5.357	6.64	31.40	5.603	6.95
26.48	5.145	6.38	28.80	5.366	6.65	31.50	5.612	6.96
26.58	5.155	6.39	28.90	5.375	6.66	31.60	5.621	6.97
26.64	5.161	6.40	29.00	5.385	6.67	31.70	5.629	6.98
26.75	5.173	6.41	29.10	5.394	6.69	31.80	5.639	6.99
26.83	5.179	6.42	29.20	5.403	6.70	31.90	5.648	7.00
26.92	5.183	6.43	29.30	5.413	6.71	32.00	5.656	7.01
27.00	5.196	6.44	29.40	5.422	6.72	32.10	5.665	7.02
27.08	5.203	6.45	29.50	5.431	6.73	32.20	5.674	7.03
27.16	5.211	6.46	29.60	5.440	6.74	32.30	5.683	7.04
27.25	5.220	6.47	29.70	5.449	6.75	32.40	5.692	7.05
27.33	5.227	6.48	29.80	5.458	6.76	32.50	5.700	7.06
27.41	5.235	6.49	29.90	5.467	6.78	32.60	5.709	7.07
27.50	5.244	6.50	30.00	5.477	6.79	32.70	5.718	7.09
27.60	5.253	6.51	30.10	5.486	6.80	32.80	5.727	7.10
27.67	5.260	6.52	30.20	5.495	6.81	32.90	5.736	7.11
27.75	5.267	6.53	30.30	5.504	6.82	33.00	5.744	7.12
27.84	5.276	6.54	30.40	5.513	6.83	33.10	5.753	7.13
27.92	5.283	6.55	30.50	5.522	6.85	33.20	5.761	7.14
28.00	5.291	6.56	30.60	5.531	6.86	33.30	5.770	7.15
28.10	5.300	6.57	30.70	5.540	6.87	33.40	5.779	7.16
28.20	5.310	6.58	30.80	5.549	6.88	33.50	5.787	7.17
28.25	5.315	6.59	30.90	5.558	6.89	33.60	5.796	7.18
28.33	5.322	6.60	31.00	5.567	6.90	33.70	5.805	8.19
28.40	5.329	6.61	31.10	5.576	6.91	33.80	5.814	8.20
28.50	5.335	6.62	31.20	5.585	6.92	33.90	5.822	8.21
28.60	5.347	6.63	31.30	5.595	6.94	34.00	5.830	8.23

EXPLANATION OF TWIST TABLE.

(See remarks on twist in Chapter V.) This table is calculated for the twist of roving on the Lowell speeder; though it may be used for other frames. It will be noticed that the multiplier used is 1.24 (see twist of one hank). It will also be noticed that the square root of the hank is worked out at each interval; so the table will be of use to all classes and "creeds" of twist. All that is necessary, after ascertaining the hank, is to look on the table for the square root of that particular hank, and then multiply it by any number that may be thought right. In this way any carder can be suited and gratify his own taste in this particular. The writer uses 1.20 for slubbers and fly frames, and 1.24 for the Lowell speeder, and 1.18 for the old soft work.

The manner of preparing this table was the following: Take 4½ hank, for an example, written decimally, and extract the square root, thus:

$$
\begin{array}{r}
2)\,\dot{4}.\dot{50}\,(2.121 \\
4 \\
\hline
41)\,50 \\
41 \\
\hline
422)\,900 \\
844 \\
\hline
4241)\,5600 \\
4241 \\
\hline
1359
\end{array}
$$

We find by this operation that the square root of 4.50 is 2.121; and by multiplying 2.121 by 1.24, we find the twist should be 2.63 per inch.

$$
\begin{array}{r}
2.121 \\
1.24 \\
\hline
8484 \\
4242 \\
2121 \\
\hline
2.63004
\end{array}
$$

It is not to be supposed that we can always find gears that will give us exactly the amount of twist we may desire; then take the gear that comes nearest as the best that can be done.

RULE TO FIND TWIST.

First, find the number of revolutions of front roll; second, its diameter; third, its circumference; fourth, multiply the revolutions of front roll by the circumference, and that gives the number of inches delivered per minute; fifth, find the number of revolutions of flyer; sixth, divide the inches delivered into revolutions of flyer, and the quotient will be the twist.

EXAMPLE.

Suppose the front roll make 120 revolutions per minute, and its diameter is 1⅛ inches (written 1.12).

$$\begin{array}{r} 3.1416 \\ 1.12 \\ \hline 62832 \\ 31416 \\ 31416 \\ \hline 3.518592 \end{array}$$

We find the circumference to be 3.518592—or, as we should write it, 3.52 inches. The revolutions of front roll 120:

$$\begin{array}{r} 3.52 \\ 120 \\ \hline 7040 \\ 352 \\ \hline 422.40 \end{array}$$

So we find that the front roller delivers 422.40 inches per minute. We will suppose the flyers run 1125 per minute.

$$\begin{array}{r} 422.40)\,1125.00\,(2.66 \\ 84480 \\ \hline 280200 \\ 253440 \\ \hline 267600 \\ 253440 \\ \hline 14160 \end{array}$$

We find in this case that the twist of a frame of the above arrangement would be 2.66 per inch.

It is probably known to carders generally that the twist gear in a slubber or fly frame is a driver, and in the Lowell speeder and some others, the twist gear is a driven. In the former more twist requires a smaller gear, in the latter a larger one.

The author has had it in mind to insert a table of diameter, area and circumference of circles, in this volume; but as there are so many such tables in existence, it is hardly advisable to do so, perhaps, but he will give the rules for obtaining them.

To find the circumference of a circle when the diameter is given, multiply the diameter by 3.1416, and the product will be the circumference. To find the diameter when the circumference is given, reverse the operation. To find the area of a circle, square the diameter (that is, multiply it into itself), and multiply that product by the decimal .7854, and the product will be the area. To find an equal square side of a circle, take the square root of the area. To find the cubical contents of a round ball, cube the diameter (that is, multiply it into itself twice: suppose the diameter is 4 inches, then 4×4=16, and 16×4=64), and multiply that product by the decimal .5236, and that product will be the solid contents of the ball.

HYGROMETER TABLE.

3¾ *44 p.c.		3⅞ *45 p.c.		4 *46 p.c.		4⅛ *47 p.c.		4¼ *48 p.c.		4⅜ *49 p.c.		4½ *50 p.c.		4⅝ *51 p.c.		4¾ *52 p.c.		5 *53 p.c.		5¼ *54 p.c.		5⅜ *55 p.c.		5½ *56 p.c.	
°		°		°		°		°		°		°		°		°		°		°		°		°	
43	93	44	93	45	93	46	93	47	93	48	93	49	93	50	93	51	93	52	93	53	93	54	93	55	93
42	87	43	87	44	87	45	87	46	87	47	87	48	87	49	87	50	87	51	87	52	87	53	87	54	87
41	81	42	81	43	81	44	81	45	81	46	81	47	81	48	81	49	81	50	81	51	81	52	81	53	81
40	76	41	76	42	76	43	76	44	76	45	76	46	76	47	76	48	76	49	76	50	76	51	76	52	76
39	71	40	71	41	71	42	71	43	71	44	71	45	71	46	71	47	71	48	71	49	71	50	71	51	71
38	66	39	66	40	66	41	66	42	66	43	66	44	66	45	66	46	66	47	66	48	66	49	66	50	66
37	62	38	62	39	62	40	62	41	62	42	62	43	62	44	62	45	62	46	62	47	62	48	62	49	62
36	58	37	58	38	58	39	58	40	58	41	58	42	58	43	58	44	58	45	58	46	58	47	58	48	58
35	54	36	54	37	54	38	54	39	54	40	54	41	54	42	54	43	54	44	54	45	54	46	54	47	54
34	51	35	51	36	51	37	51	38	51	39	51	40	51	41	51	42	51	43	51	44	51	45	51	46	51
33	48	34	48	35	48	36	48	37	48	38	48	39	48	40	48	41	48	42	48	43	48	44	48	45	48
32	45	33	45	34	45	35	45	36	45	37	45	38	45	39	45	40	45	41	45	42	45	43	45	44	45
31	42	32	42	33	42	34	42	35	42	36	42	37	42	38	42	39	42	40	42	41	42	42	42	43	42
30	39	31	39	32	39	33	39	34	39	35	39	36	39	37	39	38	39	39	39	40	39	41	39	42	39
29	36	30	36	31	36	32	36	33	36	34	36	35	36	36	36	37	36	38	36	39	36	40	36	41	36
28	33	29	33	30	33	31	33	32	33	33	33	34	33	35	33	36	33	37	33	38	33	39	33	40	33
27	30	28	30	29	30	30	30	31	30	32	30	33	30	34	30	35	30	36	30	37	30	38	30	39	30
26	27	27	27	28	27	29	27	30	27	31	27	32	27	33	27	34	27	35	27	36	27	37	27	38	27
25	24	26	24	27	24	28	24	29	24	30	24	31	24	32	24	33	24	34	24	35	24	36	24	37	24
24	21	25	21	26	21	27	21	28	21	29	21	30	21	31	21	32	21	33	21	34	21	35	21	36	21

5⅝		5⅞		6		6¼		6⅜		6½		6¾		7		7¼		7½		7¾		7⅝		8	
*57 p. c.		*58 p. c.		*59 p. c.		*60 p. c.		*61 p. c.		*62 p. c.		*63 p. c.		*64 p. c.		*65 p. c.		*66 p. c.		*67 p. c.		*68 p. c.		*69 p. c.	
o		o		o		o		o		o		o		o		o		o		o		o		o	
56	93	57	93	58	93	59	93	60	94	61	94	62	94	63	94	64	94	65	94	66	94	67	94	68	94
55	87	56	87	57	87	58	87	59	88	60	88	61	88	62	88	63	88	64	88	65	88	66	88	67	88
54	81	55	81	56	81	57	81	58	82	59	82	60	82	61	82	62	82	63	82	64	82	65	82	66	82
53	76	54	76	55	76	56	76	57	77	58	77	59	77	60	77	61	77	62	77	63	77	64	77	65	77
52	71	53	71	54	71	55	71	56	72	57	72	58	72	59	72	60	72	61	72	62	72	63	72	64	72
51	66	52	66	53	66	54	66	55	67	56	67	57	67	58	67	59	67	60	67	61	67	62	67	63	**67**
50	62	51	62	52	62	53	62	54	63	55	63	56	63	57	63	58	63	59	63	60	63	61	63	62	**63**
49	58	50	58	51	58	52	58	53	59	54	59	55	59	56	59	57	59	58	59	59	59	60	59	61	**59**
48	54	49	54	50	54	51	54	52	55	53	55	54	55	55	55	56	55	57	55	58	55	59	55	60	**55**
47	51	48	51	49	51	50	51	51	52	52	52	53	52	54	52	55	52	56	52	57	52	58	52	59	**52**
46	48	47	48	48	48	49	48	50	49	51	49	52	49	53	49	54	49	55	49	56	49	57	49	58	**49**
45	45	46	45	47	45	48	45	49	46	50	46	51	46	52	46	53	46	54	46	55	46	56	46	57	**46**
44	42	45	42	46	42	47	42	48	43	49	43	50	43	51	43	52	43	53	43	54	43	55	43	56	**43**
43	39	44	39	45	39	46	39	47	40	48	40	49	40	50	40	51	40	52	40	53	40	54	40	55	**40**
42	36	43	36	44	36	45	36	46	37	47	37	48	37	49	37	50	37	51	37	52	37	53	37	54	**37**
41	33	42	33	43	33	44	33	45	34	46	34	47	34	48	34	49	34	50	34	51	34	52	34	53	**34**
40	30	41	30	42	30	43	30	44	31	45	31	46	31	47	31	48	31	49	31	50	31	51	31	52	**31**
39	27	40	27	41	27	42	27	43	28	44	28	45	28	46	28	47	28	48	28	49	28	50	28	51	**28**
38	24	39	24	40	24	41	24	42	25	43	25	44	25	45	25	46	25	47	25	48	25	49	25	50	25
37	21	88	21	39	21	40	21	41	22	42	22	43	22	44	22	45	22	46	22	47	22	48	22	49	22

HYGROMETER TABLE—Continued.

8¼		8½		9		9¼		9½		9¾		10		10¼		10½		11		11¼		11½		12	
*70 p.c.		*71 p.c.		*72 p.c.		*73 p.c.		*74 p.c.		*75 p.c.		*76 p.c.		*77 p.c.		*78 p.c.		*79 p.c.		*80 p.c.		*81 p.c.		*82 p.c.	
°		°		°		°		°		°		°		°		°		°		°		°		°	
69	94	70	94	71	94	72	94	73	94	74	94	75	94	76	94	77	94	78	94	79	94	80	95	81	95
68	88	69	88	70	88	71	88	72	88	73	88	74	88	75	88	76	88	77	88	78	88	79	90	80	90
67	82	68	82	69	82	70	82	71	82	72	82	73	82	74	82	75	82	76	82	77	82	78	85	79	85
66	77	67	77	68	77	69	77	70	77	71	77	72	77	73	77	74	77	75	77	76	77	77	80	78	80
65	72	66	72	67	72	68	72	69	72	70	72	71	72	72	72	73	72	74	72	75	72	76	75	77	75
64	67	65	67	66	67	67	67	68	67	69	67	70	67	71	67	72	67	73	67	74	67	75	70	76	70
63	63	64	63	65	63	66	63	67	63	68	63	69	63	70	63	71	63	72	63	73	63	74	65	75	65
62	59	63	59	64	59	65	59	66	59	67	59	68	59	69	59	70	59	71	59	72	59	73	60	74	60
61	55	62	55	63	55	64	55	65	55	66	55	67	55	68	55	69	55	70	55	71	55	72	56	73	56
60	52	61	52	62	52	63	52	64	52	65	52	66	52	67	52	68	52	69	52	70	52	71	53	72	53
59	49	60	49	61	49	62	49	63	49	64	49	65	49	66	49	67	49	68	49	69	49	70	50	71	50
58	46	59	46	60	46	61	46	62	46	63	46	64	46	65	46	66	46	67	46	68	46	69	47	70	47
57	43	58	43	59	43	60	43	61	43	62	43	63	43	64	43	65	43	66	43	67	43	68	44	69	44
56	40	57	40	58	40	59	40	60	40	61	40	62	40	63	40	64	40	65	40	66	40	67	41	68	41
55	37	56	37	57	37	58	37	59	37	60	37	61	37	62	37	63	37	64	37	65	37	66	38	67	38
54	34	55	34	56	34	57	34	58	34	59	34	60	34	61	34	62	34	63	34	64	34	65	35	66	35
53	31	54	31	55	31	56	31	57	31	58	31	59	31	60	31	61	31	62	31	63	31	64	32	65	32
52	28	53	28	54	28	55	28	56	28	57	28	58	28	59	28	60	28	61	28	62	28	63	29	64	29
51	25	52	25	53	25	54	25	55	25	56	25	57	25	58	25	59	25	60	25	61	25	62	26	63	26
50	22	51	22	52	22	53	22	54	22	55	22	56	22	57	22	58	22	59	22	60	22	61	23	62	23

12⅓		12¾		13		13½		13⅝		14¼		14⅝		15		15½		15¾		16¼		16½		17	
*83 p. c.		*84 p. c.		*85 p. c.		*86 p. c.		*87 p. c.		*88 p. c.		*89 p. c.		*90 p. c.		*91 p. c.		*92 p. c.		*93 p. c.		*94 p. c.		*95 p. c.	
o		o		o		o		o		o		o		o		o		o		o		o		o	
82	95	83	95	84	95	85	95	86	95	87	95	88	95	89	95	90	95	91	95	92	95	93	95	94	95
81	90	82	90	83	90	84	90	85	90	86	90	87	90	88	90	89	90	90	90	91	90	92	90	93	90
80	85	81	85	82	85	83	85	84	85	85	85	86	85	87	85	88	85	89	85	90	85	91	85	92	85
79	80	80	80	81	80	82	80	83	80	84	80	85	80	86	80	87	80	88	80	89	80	90	80	91	80
78	75	79	75	80	75	81	75	82	75	83	75	84	75	85	75	86	75	87	75	88	75	89	75	90	75
77	70	78	70	79	70	80	70	81	70	82	70	83	70	84	70	85	70	86	70	87	70	88	70	89	70
76	65	77	65	78	65	79	65	80	65	81	65	82	65	83	65	84	65	85	65	86	65	87	65	88	65
75	60	76	60	77	60	78	60	79	60	80	60	81	60	82	60	83	60	84	60	85	60	86	60	87	60
74	56	75	56	76	56	77	56	78	56	79	56	80	56	81	56	82	56	83	56	84	56	85	56	86	56
73	53	74	53	75	53	76	53	77	53	78	53	79	53	80	53	81	53	82	53	83	53	84	53	85	53
72	50	73	50	74	50	75	50	76	50	77	50	78	50	79	50	80	50	81	50	82	50	83	50	84	50
71	47	72	47	73	47	74	47	75	47	76	47	77	47	78	47	79	47	80	47	81	47	82	47	83	47
70	44	71	44	72	44	73	44	74	44	75	44	76	44	77	44	78	44	79	44	80	44	81	44	82	44
69	41	70	41	71	41	72	41	73	41	74	41	75	41	76	41	77	41	78	41	79	41	80	41	81	41
68	38	69	38	70	38	71	38	72	38	73	38	74	38	75	38	76	38	77	38	78	38	79	38	80	38
67	35	68	35	69	35	70	35	71	35	72	35	73	35	74	35	75	35	76	35	77	35	78	35	79	35
66	32	67	32	68	32	69	32	70	32	71	32	72	32	73	32	74	32	75	32	76	32	77	32	78	32
65	29	66	29	67	29	68	29	69	29	70	29	71	29	72	29	73	29	74	29	75	29	76	29	77	29
64	26	65	26	66	26	67	26	68	26	69	26	70	26	71	26	72	26	73	26	74	26	75	26	76	26
63	23	64	23	65	23	66	23	67	23	68	23	69	23	70	23	71	23	72	23	73	23	74	23	75	23

EXPLANATION OF HYGROMETER TABLE.

This table is to be used in connection with the double-bulb thermometer, or hygrometer as such an instrument is called. When one bulb is kept wet, and the other dry, the dryness or dampness is indicated by the number of degrees between the two thermometers. The drier the atmosphere, the greater the difference between the two; the damper the atmosphere, the less difference there will be between the two. When they stand alike, then the air is perfectly saturated and begins to drop its load of moisture, which, however, does not often take place in-doors unless the building leaks. The use of this instrument and table in a carding-room is, to enable the carder to govern his weight according to the condition of the atmosphere. Cotton absorbs about five per cent. of its own weight in moisture when the air is very damp, and when the air is very dry it absorbs little or none.

It will be readily seen from the above that variations in weight of roving must be made to correspond to variations of the atmosphere in the room. If it is made to weigh no more when it has two or three per cent. of water in it, than it does when dry, as a matter of course when the yarn from such roving becomes dry it will be too light, and the cloth too light.

The author has had some knowledge of hygrometers of different kinds, and of hygrometer tables, for some years past; but there was a long time he did not understand what was meant by the terms 50, 60, or 70 per cent. of water in the air. He could read thermometers and tables, and tell, as accurately as any other man could tell, what amount of moisture there was in the air at any time, if he had the instrument and table to read from; but he did not understand the subject. Some years ago he wrote to a gentleman in Philadelphia, Pa., for some instructions in table work. In reply he received some documents which gave the desired information, and from these documents he learned how to make a table, and was enabled to understand it. Among the above-mentioned documents was a "Table of Quantity," showing the weight in grains of a cubic foot of vapor at different temperatures from 0° to 95° Fahrenheit; also, a table of corrections, and rules for calculating a table. The author wrote to the same gentleman a few weeks since, asking the privilege of introducing those tables in this work, but has received no answer, and therefore they are omitted. He does not feel at liberty to copy other men's productions without permission. He does, however, feel at liberty to give the plan on which this table was made.

It is not claimed that this table is exactly accurate. It is not brought down so fine as would probably be required for very nice experiments. It is claimed by most meteorologists that there are two degrees of absolute dryness to one observed; that is, when the two thermometers are one degree apart, it is said that there exist two degrees of absolute dryness. Some even make a little addition to that, and multiply the degrees observed by two and a small decimal for the absolute dryness. The figures on the left of double columns, designated by an asterisk (*), are for the dry-bulb thermometer; the columns marked o are for the wet-bulb thermometer; the columns marked p. c. denote the percentage of moisture in the air.

EXAMPLE.

Suppose the dry-bulb thermometer stands at 76°, and the wet at 70°; we follow down the column marked o until we come to 70°, and exactly opposite in the column marked p. c. we find 67 per cent. The figures at the top of double columns (whole and mixed numbers) denote the number of grains of water a cubic foot of air will hold at that particular temperature (dry-bulb).

THE WAY IT WAS MADE.

In the above example, 76° dry, 70° wet, difference observed 6°. In this case 6°×2.16=13° (nearly); 76°−13°=63°; and we find the air at a temperature of 76° will hold 10 grains of water, and at 63° it holds 6.75 grains. Then, by the rule of percentage, we divide:

$$10.)\ 6.75\ (.67$$
$$\underline{60}$$
$$75$$
$$\underline{70}$$
$$5$$

We find we have 67 per cent. of water in the air; that is, 67 per cent. of 10 grains—what it is capable of holding at a temperature of 76°. The wet-bulb thermometer, or its readings, is called "the dew point"; and when they both stand alike there is a hundred per cent. of water in the air, or in other words it is full, and rain begins to fall.

Several firms in Boston and some in Philadelphia make hygrometers; and Charles Wilder, of Peterboro', N. H., makes splendid instruments of this kind—beautiful in appearance, that can be relied on for accuracy. The fountains should be filled with soft water that has been boiled; the lace or silk which covers the bulb of the wet thermometer, as well as the wick that supplies it with water, should be changed occasionally. When there are only two or three degrees difference between the two thermometers, it is safe to allow one grain extra in twenty-five, in the weight of roving. The hygrometers seldom show less than 50 per cent. of water in the air in-doors. Dry-bulb thermometer 70° and wet 63° or 64° are the best conditions for carding. The variations out doors are much more; it is interesting to watch them. Almost any man of ordinary intelligence, with one of these instruments, a barometer and vane, could with a little care and attention soon become quite a respectable " Prob."

TABLE

TO ASCERTAIN THE LENGTH OF BELTING IN A ROLL, THE INSIDE COIL SUPPOSED TO BE 1½ INCHES DIAMETER.

Coils.	Feet.	Inches.	Coils.	Feet.	Inches.	Coils.	Feet.	Inches.
1		5	35	91	7	69	334	1
2		11	36	96	7	70	343	6
3	1	7	37	101	8	71	353	1
4	2	4	38	106	11	72	362	9
5	3	3	39	112	4	73	372	7
6	4	4	40	117	10	74	382	6
7	5	6	41	123	5	75	392	7
8	6	10	42	129	2	76	402	9
9	8	3	43	135	1	77	413	1
10	9	10	44	141	1	78	423	7
11	11	6	45	147	3	79	434	2
12	13	4	46	153	6	80	444	11
13	15	4	47	159	11	81	455	9
14	17	5	48	166	6	82	466	9
15	19	8	49	173	2	83	477	11
16	22	0	50	180	0	84	489	2
17	24	7	51	186	11	85	500	6
18	27	1	52	194	0	86	512	0
19	29	10	53	201	2	87	523	8
20	32	9	54	208	6	88	535	5
21	35	9	55	216	0	89	547	4
22	38	10	56	223	7	90	559	4
23	42	2	57	231	3	91	571	6
24	45	7	58	239	2	92	583	10
25	49	1	59	247	1	93	596	3
26	52	9	60	255	3	94	608	10
27	56	6	61	263	6	95	621	6
28	60	6	62	271	10	96	634	4
29	64	6	63	280	4	97	647	4
30	68	9	64	288	11	98	660	5
31	73	0	65	297	9	99	673	7
32	77	6	66	306	7	100	686	11
33	82	1	67	315	8			
34	86	9	68	324	10			

These calculations are based on the supposition that the average thickness of the belt is ¼ of an inch. If it is more, the length will run short of what the table indicates; if less, it will overrun. Count the coils, measure the diameter of the roll. If the number of coils equal two to the inch, then the table indicates correctly. If there are more than two coils to an inch of diameter, the feet will overrun what the table says; if less, it will fall short.

ILLUSTRATION OF TABLE.

We will suppose the first, or inside, coil to be $1\frac{1}{2}$ inches diameter, the circumference 4.71 inches, called in the table 5 inches. If the belt is $\frac{1}{4}$ of an inch in thickness, another coil would make the diameter of the roll 2 inches; circumference 6.28 +5=11 inches. This table was made after that manner. Suppose we have a roll of belting with 60 coils, and the diameter of the roll is 31 inches, then there is 255 feet 3 inches in the roll. It is proper to remark that 1 inch in diameter is always to be thrown out for the hole in the centre of the roll.

A PERPETUAL CALENDAR.

The author has thought that a calendar which could be used year after year, with a slight manipulation, may be of some benefit to those possessing it; and, as a common calendar for one year is of no use after that particular year has passed, and only takes the room that might be put to some use, he has concluded to place the appended perpetual calendar in this work, or one which may be made so with very little trouble.

Saturday,	1	8	15	22	29
Sunday,	2	9	16	23	30
Monday,	3	10	17	24	31
Tuesday,	4	11	18	25	
Wednesday,	5	12	19	26	
Thursday,	6	13	20	27	
Friday,	7	14	21	28	

Sunday,	1	8	15	22	29
Monday,	2	9	16	23	30
Tuesday,	3	10	17	24	31
Wednesday,	4	11	18	25	
Thursday,	5	12	19	26	
Friday,	6	13	20	27	
Saturday,	7	14	21	28	

Monday,	1	8	15	22	29
Tuesday,	2	9	16	23	30
Wednesday,	3	10	17	24	31
Thursday,	4	11	18	25	
Friday,	5	12	19	26	
Saturday,	6	13	20	27	
Sunday,	7	14	21	28	

Tuesday,	1	8	15	22	29
Wednesday,	2	9	16	23	30
Thursday,	3	10	17	24	31
Friday,	4	11	18	25	
Saturday,	5	12	19	26	
Sunday,	6	13	20	27	
Monday.	7	14	21	28	

Wednesday,	1	8	15	22	29
Thursday,	2	9	16	23	30
Friday,	3	10	17	24	31
Saturday,	4	11	18	25	
Sunday,	5	12	19	26	
Monday,	6	13	20	27	
Tuesday,	7	14	21	28	

Thursday,	1	8	15	22	29
Friday,	2	9	16	23	30
Saturday,	3	10	17	24	31
Sunday,	4	11	18	25	
Monday,	5	12	19	26	
Tuesday,	6	13	20	27	
Wednesday,	7	14	21	28	

Friday,	1	8	15	22	29
Saturday,	2	9	16	23	30
Sunday,	3	10	17	24	31
Monday,	4	11	18	25	
Tuesday,	5	12	19	26	
Wednesday,	6	13	20	27	
Thursday,	7	14	21	28	

It will be noticed that by repeating the weekly table seven times, we have the day that each month in every year must commence on for all time to come; and it is only necessary to dot the day that any month begins, with a pencil, to have a calendar for that month. The operation can be repeated during life, if desired, and then it can be left to posterity. As January, 1876, commences with Saturday, this calendar begins to reckon on that day.

ADDRESS TO A COTTON CARDER.

Friend and Brother Carder:

Some one, who was probably well qualified to judge, has had sufficient confidence in your ability and integrity to place you in charge of a very important department in cotton manufacturing. Your position is a responsible one. There is committed to your care a large amount of valuable machinery and stock, and the preparation of material for the use of all the other departments in your mill. Upon your ability and faithfulness depend the financial success of the concern for which you work, the comfort of hundreds of laboring men, women and children, and the satisfaction of thousands of people who purchase and use the goods you help to manufacture. If through your vigilance and industry your machinery is all in good order, everything well adapted to the end for which it was made, every machine and part of machine doing its appropriate work, properly attended, and producing good, smooth, even, clean roving, then the work throughout the entire mill will run well; spinners, spoolers, warpers, dressers and weavers will be able to produce good work comfortably, and good, nice cloth will be the result. On the other hand, if your cards are dull, if your drawing cuts and from any cause does not draw evenly, if through your ignorance, or negligence, or both, your machines are not well adapted one to another, or if they are not well attended, and as a result your roving is rough, uneven and dirty; then spinners, spoolers, warpers, dressers and weavers will be obliged to labor hard to produce poor work, and bad cloth will be the result.— Weary limbs, aching heads and hearts, second quality of cloth, and disappointed hopes of owners, will be the constant mementos of your inefficiency. How to avoid the latter result, and how to secure the former, should be your constant study.

An old and successful carder was once asked what was the secret of his success, and what course it was necessary for a young man to pursue, in order to attain to a high degree of perfection in this art. The old man's reply was—"*Attend to it.*" And this answer, though short, is very comprehensive. This trade requires the close application of him who would succeed in it. Success and victory do not come of ease and carelessness, but from hard study, untiring vigilance and constant attention. It makes no difference who the man is, what his attainments are, what his natural or acquired ability may be, or what his qualifications are. Hard work and close attention are the constant concomitants of this vocation; they cover all the ground in every man's experience.

Have you a good reputation as a carder? Then they are necessary for *you*, in order that you may retain it; for a love of ease and habits of carelessness will soon rob you of it, however high you may have risen. Are you a young man without a reputation? Then they are necessary for *you*, that you may thereby gain one. Did you take somebody's old room, all run down, or did you take a room all in perfect order, or did you start one exactly to your own liking? Are your tools poor and ill adapted to the place they are in, or are they all nice, exactly as you would have them? Are you working poor stock, or is it good, long staple, and clean? Have you a poor and unreliable set of help, or are they the best in the country, or are they average in quality? Is your employer strict, hard to please, sometimes unreasonable, or is he loose and careless, allowing you to do as you please? Are your associates all good men who understand their business, or are they ignoramuses who don't care a straw which way things go? Under any of these circumstances, keep at it through thick and thin; industry, perseverance, watchfulness and never-ceasing application are necessary all the time; and, what is more, these are sure to win, sooner or later. No matter what the circumstances or surroundings may be, brains, like water, may be disturbed and forced up or forced down out of their natural level, by temporary conditions, but, be sure, they will find their own level in time. Clouds, thick and black, caused by jealousy and hypocrisy, may for a while cover up talent, truth and energy; but time will clear these all away, so that they will shine out in an unmistakable manner, while the opposite will go to oblivion, where they belong.

Aim to be a better carder than any who have preceded you. It is possible for you to be so. Never be satisfied with present attainments, or think for a moment that your work is going well enough while it is possible for you to make it go better. Improve every opportunity to learn; you may learn from high and low, rich and poor, in early morning, at noon and night; you may learn from experience, from books, and from men. Make carding cotton your principal business, though it is neither necessary nor profitable to confine yourself to that alone. Most men have some one business in which they excel, and other arts and sciences are made auxiliary thereto. Few men become adepts at more than one kind of business; and in order to do that it is often necessary to make many other things contribute something towards that one. A fair education in the common branches is indispensable to proficiency in this business, and a tolerable knowledge of natural philosophy, mechanics, chemistry, meteorology, astronomy, hydraulics and hydrostatics would be better still. Set an example worthy to be imitated by your help in promptness, faithfulness, fairness, neatness and industry. Don't be a loafer yourself, and have no loafers about you. Be civil and kind to all, and especially to the poor. Be the helper of the helpless. Have good, wholesome rules and regulations; have them understood, and then see that they are enforced. It is not enough for you to give an order; it is your business to see that your orders are executed. Be respectful to superiors, and behave in such a manner yourself as to be worthy of respect from equals and inferiors. Yield strict obedience to your employer in all things, as long as you work for him. No matter how much you think you know, or how little you think he knows. If you and he don't think alike, it is your place to yield, and his to direct. You may suggest and propose in a proper manner, but don't bore him with your peculiar ideas, when you have reason to think they are distasteful to him. Let your works praise you. One thing more—please not forget it:

"Therefore all things whatsoever ye would that men should do to you, do ye even so to them." (Matt.: vii: 12.)

WEIGHTS OF VARIOUS ARTICLES.

A gallon of fresh water weighs 8 lbs., 5⅜ oz.
A gallon of oil weighs 7.45 lbs.
A gallon of vinegar weighs 8.64 lbs.
A gallon of molasses weighs 11 lbs.
Cast-iron weighs per cubic foot, 451.68 lbs.
Wrought-iron weighs per cubic foot, 486.13 lbs.
Lead weighs per cubic foot, 708.75 lbs.
A cubic inch of cast-iron weighs .2608 lb.
A cubic inch of wrought-iron weighs .2807 lb.
A cubic inch of lead weighs .4101 lb.

To ascertain the capacity in bushels of a box, bin, barrel or other vessel (dry measure), find the cubic inches, and divide them by 2150.42; the quotient is the bushels.

To find the capacity in gallons of any vessel, find the cubic inches, and divide them by 231; the quotient will be the gallons.

Water is at its greatest density at 39° Fahrenheit. It expands both ways from this point.

Ice is one-ninth more bulky than water at the above point, and steam seventeen hundred times.

RECEIPTS.

How to make glue that will hold, spread well, and keep well: Put 1 pound of good, dry glue to soak in good cider vinegar enough to cover it; let it soak over night. In the morning put the glue pot heating over a slow fire (sitting in a kettle of water is best). When it is pretty well melted, put in one gill of molasses; and after it is all dissolved, put in 100 grains of borax and 100 grains of pulverized chalk. The borax will melt, and the chalk sink. Thin with water to the right consistency.

How to make belt cement: Dissolve fish glue (or, as some call it, isinglass) in rum and water half-and-half. It dissolves quickly, is ready to use as soon as dissolved, and is best when first made. Make no more at any time than is wanted for immediate use.

How to make a roll varnish for top rollers that are old and dry, or rough: Mix common shellac and raw linseed oil—two parts of the former to one of the latter. Sand-paper the rolls first, and then put the varnish on cold. Rub it in well with a piece of yarn waste or a piece of cloth.

Another nice preparation for top rollers may be made of gum arabic and gelatine, half-and-half, dissolved in water and rubbed on with the finger. If rollers lap up, after receiving a coat of shellac and oil, then put on a little of this last-named varnish, and they will be all right.

POWER.

The mechanical powers are three in number, according to Haswell, viz: the lever, the inclined plane and the pulley. Carders have most to do with the lever, as that kind of weight is commonly used in connection with most machines. To find the power, when the length of the lever and the weight are known, multiply the weight by the distance from the fulcrum, and divide that product by the distance of the fulcrum from the power.

EXAMPLE.

A drawing weight weighs 6 pounds. It hangs on a lever $6\frac{1}{2}$ inches from the fulcrum, and the fulcrum is $1\frac{1}{2}$ inches from the power. What is the power on the rollers?

```
        6.50
           6
      ───────
 1.50) 39.00 (26 lbs., Ans.
        300
      ───────
         900
         900
```

RULE TO CALCULATE DRAUGHT.

A few remarks on draught may be in order before giving the rule to calculate it, as wrong ideas are often given and received by the use of common terms. The meaning of the term as used in connection with machinery for carding and spinning cotton, is to draw out, stretch or elongate a sliver of cotton. There is no misunderstanding of terms when we say a draught of 3, 4, 6, 7 or $7\tfrac{1}{2}$. It means that for every inch that a frame receives, it delivers 3, 4, 6, 7 or $7\tfrac{1}{2}$ inches. Misunderstandings occur where the draught is less than 1, as when we say a draught of $1\tfrac{1}{4}$ or $1\tfrac{1}{2}$.

If a machine delivers exactly what it receives, of course there is no draught or stretch; but according to rule, all the rolls being alike in diameter, and all the gears, drivers and drivens, the same figures would say there was a draught of 1 because the drivens would contain the drivers one time exactly. It will be understood by a careful study of this subject, and especially the rule, that one is in all cases to be deducted from draught as shown by figures; the remainder is the real draught or stretch; so that when we say a draught of $1\tfrac{1}{8}$, $1\tfrac{1}{4}$, or $1\tfrac{1}{2}$, it is really a draught of $\tfrac{1}{8}$, $\tfrac{1}{4}$, or $\tfrac{1}{2}$. The author makes these remarks because he has been puzzled himself over these very things, and he has never seen this subject referred to in any work on cotton manufacturing.

The place to begin to count teeth, and take measures for calculating the draught of any machine, is at its delivery, and work back to the receiving roll. If the whole draught of a machine is sought, intermediate rolls, cylinders, or gears need not be taken into account, unless they are directly connected by way of driving. The middle rollers of a drawing, roving, spinning frame or the doffer or feed rollers of a card, need not be reckoned, unless in case of a card where there are no calender rollers; in that case the doffer is the delivering roller and must be so reckoned. We sometimes find a gear which is both a driver and a driven. In such a case leave it out, by the rule of cancellation. Call the gear connected with the delivering roll the first driver in all cases (whether it is really a driver or a driven gear), and the gear connected with the receiving roller is to be reckoned as the last driven always.

The principle of draught is, to run a sliver through a succession of rollers—each pair, or set, as it advances, running a little faster than the preceding—so that the rule to find draught is based on the plan of an increase of speed of rollers; and if both receiving and delivering rollers were equal in size, the question would amount to this: How many revolutions does this delivering roller make to the receiving roller one? But as there is most always a difference in size between the receiving and delivering rollers, the question actually is: How many inches are delivered for one received? And the difference in size of rollers must be taken into account.

How to proceed to take draught.— First, pick out all the driving gears, beginning as directed above; count the teeth of each, and place the number of teeth, or the diameter in inches of every gear, or pulley, above a line, thus:

1st driver, 44 teeth; 2nd, 50; 3rd, diam. receiving roll $\frac{8}{8}$
___
1st driven, 108 " ; " 77; "   " delivering " $\frac{10}{8}$

Second, count the drivens in the same way, and place them under the line. We suppose in this case that the front or delivering roller is $1\frac{1}{4}$ inches, and the back or receiving roller is 1¼ inches, which are equal to ⅜ and ¹⁰⁄₈ in the shape of an improper fraction. We next multiply all the drivers and the diameter of the receiving roller together: 44×50×9=19,800. Next we multiply all the drivens and the diameter of the delivering roller together: 108×77×10=83,160. Next divide the product of drivens by the product of drivers; the quotient is the draught sought.

$$19800 \overline{)83160} (4.20$$
$$\underline{79200}$$
$$39600$$
$$\underline{39600}$$

We may sum up the rule as follows:

I. Multiply all the drivers together, and that product by the diameter of the receiving rolls.

II. Multiply all the drivens together, and that product by the diameter of the delivering rolls.

III. Divide the product of the drivens by the product of the drivers; the quotient is the draught.

NOTE.—If the frame is driven by pulleys instead of gears, or if there are both pulleys and gears, take the diameter of pulleys in inches and decimal parts of inches, instead of teeth.

RULE TO CALCULATE SPEED.

The principle of calculating speed is the same as calculating draught, with this difference:—Draught has to do with the surface, and speed only with the revolutions.

The speed of different drums, pulleys, or shafting, is in proportion to the diameter, or it may be calculated after that manner. Suppose a driving pulley or drum is two feet in diameter, and it drives a drum or pulley one foot in diameter; then the latter makes two revolutions to the former one, and so of a succession of drums or pulleys, no matter how many; so that a succession of such, or the speed, may be calculated one way by multiplication, and the reverse by division. Suppose a driving pulley or drum is twelve feet in diameter, and it belts on to a pulley five feet in diameter, and this latter shaft contains a second driving pulley four and a half feet in diameter and belts on to a driven three feet in diameter. The speed of the main driver is one hundred and fifty revolutions per minute. What would be the speed of the last mentioned shaft or drum?

          Inches.   Inches.
1st driver=144; 2nd driver=54; speed of main driver, 150

          Inches.   Inches.
1st driven=60; " driven=36

In this example we first multiply the drivers together, and that product by the speed of main driver. Next we multiply the drivens together, and take the product for a divisor for the product of the drivers.

```
          144             36
           54             60
         ———            ————
          576           2160
          720
         ————
         7776
          150
       ——————
       388800
         7776
      ————————
  2160 ) 1166400 ( 540
         10800
         —————
           8640
           8640
          —————
             00
```

We find by this operation that the speed of the last-named drum is 540 revolutions per minute. Now, suppose we have a shaft driven in the manner described, the speed of which is 540 revolutions per minute, and we wish to ascertain the diameter of the main driver, we reverse the operation: $2160 \times 540 = 1166400 \div 140 = 7776 \div 54 = 144$ inches.

From these examples and illustrations we derive the following general rules for calculating speed:

I. Multiply all the drivers (diams.) together, and that product by the speed of the main driver.

II. Multiply all the drivens together.

III. Take the products of the drivens for a divisor for the product of the drivers; the quotient will be the speed sought.

NOTE. With gears proceed in the same manner, counting teeth instead of taking diameter.

TIME.

Every carder should take note of time as it passes, so that he may realize how fast it goes, and act with promptness in the discharge of his duties. It is interesting, if not profitable, to notice its divisions and variations. The sun is the great source of light; it is supposed to stand still; but the motion of the earth around it gives us four seasons: twelve calendar months, fifty-two weeks, three hundred sixty-five and a fourth days, eight thousand seven hundred and sixty-six hours, five hundred twenty-five thousand nine hundred and sixty minutes, thirty-one million five hundred fifty-seven thousand and six hundred seconds, in round numbers, in a year.

The average length of sunlight for the year is not far from twelve hours a day, though the extremes go as high as fifteen hours and sixteen minutes, and as low as nine hours and three minutes. We have clocks and watches to keep time, and artificial light to help equalize the days; but the sun and the motions of the earth during the year are the great regulators. When the sun is south we call it noon, but scientific men have discovered a little variation here, and fortunately for us, they have told how much, and how we may keep regular time during these variations. The author has come upon a nice little table in a certain work which gives the exact time, by the clock, when the shadow is at the noon mark, for every day in the year. He wrote to the proper persons and asked liberty to make an extract from their table, but having received no answer, he will make a little calculation by the Old Farmers' Almanac, showing what time it is, or should be, by the clock when the sun reaches the noon mark, not for every day in the year, but often enough to enable us to regulate a clock. It will be noticed that the sun is south four days in the year only, viz: April 15th, June 15th, September 2nd and December 24th.

A reference to the following table will show what time it should be when the sun reaches the noon mark on the following days in the year:

WHEN THE SUN REACHES THE NOON MARK.

DATE.	True time, M.	DATE.	True time, M.	DATE.	True time, M.	DATE.	True time, M.
Jan.	H. M	*April.*	H. M.	*July.*	H. M.	*Oct.*	H. M.
1	12.04	1	12.04	2	12.04	3	11.49
5	12.06	5	12.03	6	12.05	6	11.48
10	12.08	10	12.01	14	12.06	10	11.47
15	12.10	15	12.00			15	11.46
20	12.11	20	11.59	*Aug.*		20	11.45
25	12.13	25	11.58	7	12.05	25	11.44
30	12.14	30	11.57	15	12.04		
				20	12.03	*Nov.*	
Feb.		*May.*		25	12.02	13	11.45
5	12.14	5	11.56	29	12.01	20	11.46
20	12.14	23	11.57			25	11.47
25	12.13	31	11.58	*Sept.*		28	11.48
				2	12.00	30	11.49
Mch.		*June.*		4	11.59		
1	12.12	7	11.59	7	11.58	*Dec.*	
6	12.11	15	12.00	10	11.57	3	11.50
10	12.10	21	12.02	12	11.56	5	11.51
15	12.09	28	12.03	15	11.55	10	11.53
20	12.07			18	11.54	15	11.56
25	12.06			22	11.52	20	11.58
				25	11.51	22	11.59
				29	11.50	24	12.00
						26	12.01
						28	12.02

To make a noon mark, find a door or window frame that stands perpendicular, then with the aid of a compass strike a line on the floor due north from the corner where the sun shines.

In response to a circular announcing the intention of the author to publish the Cotton Carders Companion, orders were received from the following individuals and firms:

Adams, Phinehas, Stark Mills, Manchester, N. H.
Atkinson, Edward, Indian Orchard Mill, Boston, Mass.
Atwood, J. S., Wauregan Mills, Wauregan, Conn.
Aspinwall, M., Attawaugan Mills, Killingly, Conn.
Ainsworth, Wm. H., Wilmington Cotton Mills, Wilmington, N. C.
Adams, Charles E., Lowell, Mass.

Battles, F. F., Massachusetts Cotton Mills, Lowell, Mass.
Burke, William A., Lowell, Mass.
Blaisdell, Wm. F., Merrimack Manufacturing Co., Lowell, Mass.
Blaisdell, C. E., Merrimack Manufacturing Co., Lowell, Mass.
Brown, H. H., Fisherville, N. H.
Bailey, C. R., Waltham, Mass.
Boyden, E., Prescott Mills, Lowell, Mass.
Bedlow, George W., Chicopee, Mass.
Bacon, Francis E., & Co., Boston, Mass.
Burnham, William J., Lewiston, Maine.

Cumnock, A. G., Boott Cotton Mills, Lowell, Mass.
Cheney, J. L., Lowell, Mass.
Clark, C. Martin, Merrimack Manufacturing Co., Lowell, Mass.
Chapin, Samuel, Tremont and Suffolk Mills, Lowell, Mass.
Cooledge, F. S., Lawrence Manufacturing Co., Lowell, Mass.
Call, A. F., Lowell, Mass.
Conihe, William, Merrimack Manufacturing Co., Lowell, Mass.
Clough, Henry, Tremont and Suffolk Mills, Lowell, Mass.
Clark, Jeremiah, Lowell, Mass.
Cumnock, J., Great Falls, N. H.
Chase, Jonathan, Albion Co., 17 Exchange St., Providence, R. I.
Chandler, Geo. H., Merrimack Manufacturing Co., Lowell, Mass.
Carroll, A. H., Mont Vernon Mills, Baltimore, Md.
Collins, David M., Merrimack Manufacturing Co., Lowell, Mass.
Caverly, Robert B., Lowell, Mass.

Dupee, James A., Boston, Mass.
Dix, Daniel, Merrimack Manufacturing Co., Lowell, Mass.
Dickey, A. J., Manchester, N. H.
Davis, A. A., Hamilton Manufacturing Co., Lowell, Mass.
Davis, S. H., Merrimack Manufacturing Co., Lowell, Mass.
Draper, George, & Son, Hopedale, Mass.
Davis, E. S., Lewiston, Maine.
Drew, W. S., Atlantic Manufacturing Co., Lawrence, Mass.
Davis, Edgar A., Nashua, N. H.
Dyer, D. H., Wilmington Cotton Mills, Wilmington, N. C.
Donaldson, R. I., English Manufacturing Co., Columbia, S. C.

Edwards, Jabez, Merrimack Manufacturing Co., Lowell, Mass.
Earle, D. D., Forestdale, R. I.

Foss & Pevey, Lowell, Mass.
Francis, James, Lowell, Mass.

Green, Benjamin, Cabot Manufacturing Co., Brunswick, Maine.
Green, Amos, Merrimack Manufacturing Co., Lowell, Mass.
Giles, Melden, Tremont and Suffolk Mills, Lowell, Mass.
Gegenheimer, E. D., Taunton, Mass.
Greenleaf, Nath'l, Merrimack Manufacturing Co., Lowell, Mass.

Hinckley, Isaac, Philadelphia, Penn.
Howe, J. S., Merrimack Manufacturing Co., Lowell, Mass.
Harding, David M., Merrimack Manufacturing Co., Lowell, Mass.
Harper, O. M., Pittsburg, Penn.
Hardy Machine Company, Biddeford, Maine.
Howard, B. B., Cordis Mills, Millbury, Mass.
Hill, William H., Dover, N. H.
Hale, F. W., South Berwick, Maine.
Haskins, S. W., Woonsocket, R. I.
Hayden, E. B., Plymouth W. and C. Factory, Plymouth, Mass.
Hardwick, Wm., Wilmington Cotton Mills, Wilmington, N. C.
Huse, H. E., Lowell, Mass.

Jordan, Charles P., West Boylston, Mass.

Kent, Hervey, Exeter, N. H.
Kitson Machine Company, Lowell, Mass.
Kimball, L. S., Lowell, Mass.
Kimball, H. C., Blackstone, Mass.
Kimball, Wilson, Merrimack Manufacturing Co., Lowell, Mass.

Ludlam, J. S., Merrimack Manufacturing Co., Lowell, Mass.
Lovejoy, S. A., Melrose, Mass.
Lowell Card Company, Lowell, Mass.
Lyford, A. B., Merrimack Manufacturing Co., Lowell, Mass.
Leonard, James, Merrimack Manufacturing Co., Lowell, Mass.
Libbey, William F., Merrimack Manufacturing Co., Lowell, Mass.
Lowrie, George M., West Boylston, Mass.
Lovering, O. W., Lyman Mills, Holyoke, Mass.
Leigh, F. A., Boston, Mass.
Lawry, F. B., Lewiston, Maine.

Moulton, O. H., Hamilton Manufacturing Co., Lowell, Mass.
Maxfield, J. P., Tremont and Suffolk Mills, Lowell, Mass.
Mason, A. A., Biddeford, Maine.
Mahoney, Jeremiah, Chicopee, Mass.
Maxfield, R. A., Nashua Manufacturing Co., Nashua, N. H.
Moody, George A., West Peterboro', N. H.
McCrillis, C. H., Wait's River, Vt.
Miles, William H., Lewiston, Maine.

Ober, Oliver, Merrimack Manufacturing Co., Lowell, Mass.

Palfrey, J. C., Boston, Mass.
Putnam, Frank P., Lowell, Mass.
Pray, Thomas, Jr., Providence, R. I.
Peach, Charles S., North Adams, Mass.
Prest, J. E., Pownal, Vt.

Sawyer, J. H., Appleton Co., Lowell, Mass.
Saunders, Benjamin, Vale Mills, Nashua, N. H.
Stearns, Charles, Merrimack Manufacturing Co., Lowell, Mass.
Smith, John W., Merrimack Manufacturing Co., Lowell, Mass.
Stiles, W. A., Tremont and Suffolk Mills, Lowell, Mass.
Symonds, Benjamin, Boott Cotton Mills, Lowell, Mass.
Straw, E. A., Amoskeag Manufacturing Co., Manchester, N. H.
Saunders, Francis, Hamilton Manufacturing Co., Lowell, Mass.
Stone, William, Merrimack Manufacturing Co., Lowell, Mass.
Slater, A. C., Whitinsville, Mass.
Sweet, Charles, Jr., Attawaugan Mills, Killingly, Conn.
Sands, B., Lewiston, Maine.
Spofford, Chas., Merrimack Manufacturing Co., Lowell, Mass.
Stevens, William, Lewiston, Maine.
Shaw, Thomas S., Tremont and Suffolk Mills, Lowell, Mass.
Scribner, George F., Lowell Machine Shop, Lowell, Mass.
Sargent, J. L., Lowell, Mass.

Thyng, C. C., Warren, R. I.
Towne, Jacob A., Manchester, N. H.
Thompson, John B., Salmon Falls, N. H.
Trueworthy, J. B., Lowell, Mass.
Tripp, John, Lowell, Mass.

Whiting, Phineas, Lowell, Mass.
Wilder, Charles, Peterboro', N. H.
Wright, Wm. A., Merrimack Manufacturing Co., Lowell, Mass.
Welch, Willard C., Merrimack Manufacturing Co., Lowell, Mass.
Wood, Horace J., Fall River, Mass.
Whitehead & Atherton, Lowell, Mass.
Watson, A. B., Lewiston, Maine.
Whitney, William S., Lowell, Mass.
Wattles, T. B., Chicopee Falls, Mass.
Weeks, George W., Lancaster Mills, Clinton, Mass.
Wright, Bliss & Faybian, Boston, Mass.
Whitney, William H. H., Boston, Mass.
Williston, A. L., Greenville Manufacturing Co., Florence, Mass.
Welch, John J., Wilmington Cotton Mills, Wilmington, N. C.
Winsor, E. M., Providence, R. I.
West, H. F., Gloucester City, N. J.
Watson, James, Merrimack Manufacturing Co., Lowell, Mass.

Cotton Carders' Companion is for sale by FOSTER WILSON, 22 Merrimack Corporation, Lowell, Mass. Sent to any part of the country, post paid, on receipt of price, $1.75.

EDWARDS'
IMPROVED CONE BELT GUIDE.

This appendage was invented and applied to the *Lowell* Speeder by Mr. Jabez Edwards, but is equally applicable to other roving frames. The design is to change the position of the belt on the cones and at the same time change the rate of let-off, as may be required to regulate the winding of roving to any kind of weather. It is simple in its construction, durable, easily managed, operates immediately, and can be changed in one-tenth of the time it takes to change a gear. Mr. Wilson has referred to it in his chapter on Speeders, in the "Cotton Carders' Companion," and has given something of a description of it, which please see. The attention of cotton manufacturers and machine builders is invited to it. It can be seen in use in the carding rooms of the Merrimack Manufacturing Company, and the inventor will be pleased to show it, and explain its principles to any one who may favor him with a call, or he will answer communications with reference to it. Address,

JABEZ EDWARDS, Lowell, Mass.

Under-Flat Cotton Card.

ADVANTAGES GAINED

With the use of the Under-Flat Cotton Card.

40 per cent. of labor,
50   "   "   floor room.
50   "   "   power,
50   "   "   belts,
50   "   "   oil,
33   "   "   clothing.

Send for circular. **FOSS & PEVEY,**
Howe St., Lowell, Mass.

JEREMIAH CLARK'S
Machinery Agency,

For the Purchase, Sale and Exchange of

All Kinds of Machinery,

New and Second-Hand.

Has on hand a Large Supply of Good Second-hand

Cotton and Woolen Machinery,

New and Second-Hand Card Clothing, Belting, Bobbins, Shuttles, Loom Pickers, etc. Also, Manufacturers' Supplies.

106 and 108 Middle Street, Lowell, Mass.

L. S. KIMBALL,
ROLL COVERER,

And Manufacturer of

Leather Loom Pickers and Card Leather Belting.

The best of material used in roll covering. All orders promptly attended to, personal attention given to work, and satisfaction guaranteed in all cases.

27 Shattuck Street, - - - Lowell, Mass.
(Lowell Card Co.'s Building.)

LOWELL CARD COMPANY,

Manufacturers of

CARD CLOTHING

OF EVERY DESCRIPTION,

Set in Leather, Rubber, and English and American Card Cloth.

Particular attention given to NAPPING CARDS, and Cards designed for Worsted work. The best of stock used, and satisfaction guaranteed in all cases.

127 Market St., Lowell, Mass.

L. L. PERRIN, President.   S. S. GOULD, Treasurer,

KITSON MACHINE COMPANY, LOWELL, MASS.,

BUILDERS OF

Patent Cotton Openers and Lappers,

With recent Valuable Improvements.

PATENT COMPOUND

OPENER LAPPER.

Shoddy and Waste Machines,

AND

RAG DUSTERS.

NEEDLE-POINTED CARD CLOTHING.

RICHARD KITSON, President.
SAMUEL E. STOTT, Treasurer.

HARDY MACHINE CO.,

Proprietors and Manufacturers of the "Hardy" Improved

Traverse Wheel Card Grinders,

For Grinding Cotton, Woolen, Linen or Silk Cards.

Pat. Oct. 24, 1865; Dec. 15, 1866; April 16, 1867; June 21, 1870. Extension granted Feb. 5, 1875.

This cut represents 'Hardy's" Improved Grinder, grinding the Doffer and Main Cylinder, one or both, at the same time. (With adjustable stands, complete.) Over 5000 of these machines are in constant use. Grinders for Top-Flats, Lickerins, Workers, Strippers, etc., 1 to 4 at one time. The grinding wheel passes its whole width beyond the range of card teeth before it changes, thus grinding all the teeth alike. Also, build machines for grinding Shears, one for trueing Dresser Rolls, and a machine for trueing Calender Rolls without removal from frame. Also, planer knife and tool grinder, to grind veneer cutters, curriers' knives, paper cutters or any long knives.

The above cut represents our machine for grinding 1 to 4 top-flats at one time, with cylinder brush for cleaning out the teeth before grinding—done perfectly. Patented April 16, 1867. 700 now in use.

HARDY MACHINE CO., Biddeford, Me.

Chas. Hardy, Treas. and Agt.

Foster Wilson, Jr.,

DEALER IN

GENTLEMEN'S

HATS, GLOVES,

AND

Furnishing Goods.

Hats Made to Order

AND

Silk Hats made Over.

ONE PRICE ONLY.

53 Central Street,

LOWELL, MASS.

www.ingramcontent.com/pod-product-compliance
Lightning Source LLC
Chambersburg PA
CBHW020309170426
43202CB00008B/556